Climate Crises in Human History

Climate Crises in Human History

Editors

A. Bruce Mainwaring
Robert Giegengack
Claudio Vita-Finzi

AMERICAN PHILOSOPHICAL SOCIETY • 2010

Lightning Rod
PRESS

AMERICAN PHILOSOPHICAL SOCIETY
Held at Philadelphia
For Promoting Useful Knowledge
Volume 6

Special Transactions #1
Published in association with the Transactions series
of the American Philosophical Society

Library of Congress Cataloging-in-Publication Data

Climate crises in human history / editors, A. Bruce Mainwaring,
Robert Giegengack, Claudio Vita-Finzi.
p. cm.
Includes bibliographical references and index.
ISBN 978-1-60618-921-4
1. Climatic changes. 2. Climatic changes—Environmental aspects.
3. Environmental management. 4. Environmental protection.
I. Mainwaring, A. Bruce., 1927– II. Giegengack, Robert Francis, 1938–
III. Vita-Finzi, Claudio.
QC903.C564 2010
904'.5—dc22
2010029252

Compositor: BookComp, Inc.
Printer: Hamilton Printing Company

The cover illustration shows an excavation in progress at Tell es-Sweyhat, an early
Bronze Age site in modern Syria whose abandonment is sometimes attributed to the
onset of arid conditions. Yet it is now clear that climate events have not had the same
effect throughout northern Syria, and that at Tell es-Sweyhat increasing aridity may
simply have encouraged a growing emphasis on pastoralism. Moreover, much of the
environmental degradation in its environs is apparently owed to human activities.
(*Photo courtesy of Michael Danti*).

Contents

Acknowledgments

It is a privilege to contribute to American Philosophical Society publications and gratifying to find that the first issue of the Transactions (1771), its oldest series, included a paper on the Change of Climate in the Middle Colonies of North America.

We thank David R. Harris, Neil Roberts, and several anonymous referees for their reviews and encouragement, the Mainwaring Archive Foundation and the University Museum for financial support, and the contributors to the discussions who are not represented in these pages. Tena Thomason, Rachelle Kaspin, Amanda Mitchell-Boyask, Pam Kosty, Margaret Spencer, and Elaine Wilner of the Museum Staff were largely responsible for the smooth and effective operation of the conference from which this volume emerged. Mary McDonald of the American Philosophical Society has provided us with valuable editorial guidance.

ABM, RG, & CVF
Philadelphia, 2010

Introduction

Climate Crises in Human History

THE EDITORS

The response of societies to severe climate change is a field of study that can no longer be dismissed as an academic exercise. Current concern over global warming is an important reason, but by no means the only one. Archaeologists and historians have been confronted for many decades with evidence that sites of human habitation were abandoned precipitously, or that entire communities experienced famine or mass disease, in the absence of self-evident explanations such as a volcanic eruption or a murderous invasion. Progress in environmental reconstruction and direct dating of archaeological materials has lent credibility to climatic solutions. In addition, a wealth of popular books and films on such themes, amateur participation in archaeological excavations, and inexpensive travel to locations rich in spectacular ruins have encouraged colorful speculation at the expense of prosaic narrative.

The respect accorded to climate hypotheses is also subject to fashion. In the early twentieth century, environmental explanations for the rise and fall of cultures, and for episodes of invasion and conflict, were much in vogue. The reaction against determinism, doubtless coloured by revulsion against racism and the emergence of the science of sociobiology, was just as extreme. The current trend is to explain the decay and collapse of the Maya culture, the departure of the sculptors of the Easter Island megaliths, and the disappearance of Norse communities by more subtle theses in which environmental stress and human failure are complementary factors.

The Museum of Archaeology and Anthropology at the University of Pennsylvania in Philadelphia has been running a series of programs on human response to extreme environmental events. These programs have addressed a range of environmental events, from meteoritic impact and extreme floods to rapid climate change. In 2008, the Museum convened a two-day conference to consider the response of selected cultures to climate events that could be documented from the archaeological or geologic records, or both. The fifteen participants in the conference were selected to cover a wide range of cultures, timescales, and analytical techniques. The coverage represented by the

community of scholars who participated is not comprehensive, but we hope its range provides compensation.

In Part I of this book, we have tried to promote this eclectic approach by focusing on the field evidence at selected sites in the belief that many key case studies were overdue for reassessment.

In Chapter 1, Giegengack and Vita-Finzi review the evidence that documents the warming trend now under way. Whereas there is no shortage of readable texts on that trend and its extrapolation into future decades, notably the successive editions of the IPCC, the scientific basis for those predictions remains fragile and in need of uninhibited discussion. The authors endorse the need for reduced reliance on fossil fuels, but they also emphasize the need to develop both mitigation and adaptation strategies based on the best data available. In their view, ill-founded panic is as dangerous as complacent inaction.

There follow four chapters that show that the record of past climates is open to competing explanations. In Chapter 2, Manning considers the problems that arise in the calibration of radiocarbon dates, the solar and climatic factors that may account for variations in ^{14}C production, and the distortion of the climatic and archaeological records that sometimes results. The content of these records should in itself be compelling if not unimpeachable. Bahn shows in Chapter 3 that something as simple as the identification of animals from rock art is by no means straightforward, and that assessing the environmental significance of such art can be equally problematic. The fact that much rock art is dated on the basis of style further hampers its value to climate history. Hodges presents a veil of volcanic dust in AD 536 as a plausible explanation for a series of events that affected millions of peasants in western Europe (Chapter 4); other interpreters of these events are drawn (in his phrase) to climatic catastrophes and events such as plagues like "moths to a flame."

Fiedel reviews in Chapter 5 the case for a bolide impact as the primary cause of both the sudden extinction of the many genera of large mammals that occupied North America during the Pleistocene and the brief return to glacial conditions known as the Younger Dryas period. The case remains contentious precisely because the evidence is often ambiguous (it often points to a climate event rather than a meteorite impact as trigger), no less ambiguous than the evidence that supports the conclusion that the impact of a large meteorite brought about extinction of the dinosaurs (and many other life forms) at the boundary of the Cretaceous and Tertiary.

The chapters in Part II explore the processes whereby groups of humans respond to a climate crisis. Vita-Finzi shows in Chapter 6 how quickly a climate effect can change from hostile to benevolent in the light of technological progress or, in the two instances he discusses, a switch in river behavior from silting to incision. The resulting challenges may well have governed flood farming in the Old World as in the New.

Drought is perhaps marginally more forgiving than encroaching ice, and the Sahara (Smith in Chapter 7) and Mesopotamia (Danti in Chapter 8) provide, respectively, prehistoric and protohistoric illustrations of the strategic

benefits of mobility (and one might add, empty space) when trying to find a reliable water supply or to maximize scanty rainfall.

In Part III, our contributors address the human element, which is discussed at some length in the earlier chapters but assumes center stage as Scarborough and Burnside consider culture to be a key variable. Drawing on their studies of the Maya and the living Balinese, in Chapter 9 they conclude that no cultural pathway is immune to catastrophe. They also demonstrate the variety of responses to a particular environment (here a semitropical one) that could, perhaps, commit the society to eventual survival or failure. Hammond discusses the Classic Maya Collapse in some detail in Chapter 10, and shows how analyses and interpretations continually change in the light of fresh data and ways of thinking; he suggests that political, demographic, and environmental explanations cannot be separated. In Chapter 11, Mormina and Higham use the techniques of genetics to investigate how and when populations in southeast Asia responded to sea-level changes driven by climate. Here the complications introduced by culture and economy are kept in check even though they evidently played a part in deciding the map and calendar of dispersal.

Scarborough and Burnside suggest that truly learning from the past might enable us to avoid the worst effects of a looming climate crisis. The editors' ambition is more modest: to highlight the uncertainties in both stimulus and response, and in so doing to rein in those who think they know what is best for the rest of us.

PART I
Evidence

1

Climate Change: Past, Present, and Future

ROBERT GIEGENGACK AND CLAUDIO VITA-FINZI

INTRODUCTION

The history of the human species has played against constantly changing conditions that have promoted both the biological evolution of our species and the emergence of the complex social structure that supports human communities. These changes have taken many forms, but none so directly affects the welfare of human communities as variation in the physical conditions with which we live on a daily basis—the weather that swirls around us, and the longer term variation in the means and extremes of that weather, the phenomenon we call climate. In this chapter we undertake to show that the nature and causes of climate change are many and complex, so that assessing the impact of climate change on human societies remains a challenging enterprise. But this complexity makes it all the more important to make those assessments as free as possible from any prejudices that interfere with successful analysis of the human contribution to current environmental trends.

The global human population has reached 6.8 billion, and it is stressing the natural systems that support us to an extent not previously realized. To reach that population, human society has learned to extract resources from Earth-surface systems at a rate that has accelerated as human society has grown more complex. That rate of extraction is clearly not sustainable, at the current size of the human population, let alone the size to which it is projected to grow over the next century, and well-documented climate change already exacerbates this problem.

THE INTERGOVERNMENTAL PANEL ON CLIMATE CHANGE

The Intergovernmental Panel on Climate Change (IPCC) was convened in 1988 to "assess on a comprehensive, objective, open and transparent basis the latest scientific, technical and socio-economic literature produced worldwide relevant to the understanding of the risk of human-induced climate" (IPCC

2010). IPCC has undertaken, over the last twenty-one years, through a series of summary publications (1990, 1995, 2001, 2007), to compile the literature of climate change, and to use the power of modern supercomputers to extrapolate trends revealed in that literature into an uncertain future.

Given that the mission of IPCC is focused on human-induced changes in climate, that body has not addressed the history of climate prior to human alteration of the chemistry of the atmosphere, other than in summary fashion (Jansen et al. 2007). The IPCC has documented the instrumental record of a rise in global mean temperature (Fig. 1) from data collected at meteorological stations throughout the world; the recent history of changes in the mass of glaciers on land (Fig. 2); the evolution of changes in sea level, as documented from tide gauges and recently launched orbiting satellites (Fig. 3); recent changes in global snow cover; and the recent history of changes in ice cover over the Arctic Ocean. The IPCC has addressed the controversial question of whether human-induced climate change has affected the intensity and/or frequency of atmospheric disturbances. The IPCC has also compiled anecdotal reports of climate change from a wide variety of terrestrial and marine environments, the great majority of which further document the warming trend now under way.

An important climate indicator, the record of rapid retreat of high-altitude glaciers at lower latitudes, has been exhaustively documented by Thompson and his colleagues (Thompson et al. 2006). From these records it is clear that the last 190 years have seen a systematic rise of global mean temperature of the order of $1°$ C; it has been larger at high northern latitudes, but the record from those localities is shorter.

The IPCC also undertakes to review the data of the changing chemistry of the atmosphere, as monitored since 1958 from an observatory on Mauna Loa, Hawaii, and now from many other sites worldwide (Fig. 4). Those data continue to show a systematically increasing concentration of atmospheric CO_2 —from 315 ppm in 1958 to 385 ppm in 2009. That increase has been attributed to the global industrial production of CO_2 from combustion of fossil fuels, roasting of $CaCO_3$ in the manufacture of cement, and land-use changes that deliver CO_2 to the atmosphere from reservoirs in soil or standing vegetation.

These figures are known with variable accuracy. In comparison with the annual global conversion of hydrocarbon fuels to CO_2 or the contribution from limestone roasting, the balance of CO_2 delivered to the atmosphere from land-use changes—to the extent that it also includes the operation of carbon sinks represented by land removed from agriculture now reverting to other mixes—has proven more vexing to quantify. However, it is clear that the atmospheric inventory of CO_2 has risen at a rate that is roughly half the rate at which industrial CO_2 has been delivered to that reservoir: the amount of carbon in the atmosphere now increases by 4.1 Gt/yr, while 7.7 Gt/yr are transferred directly from fossil-fuel reservoirs to the atmosphere (Global Carbon Budget Consortium 2009). Many research exercises are now under way to describe and quantify the processes whereby half of the anthropogenically generated CO_2 is apparently sinking into other reservoirs, and to discover how long this sequestration process is likely to prevail. The IPCC has asserted (IPCC 1990, 1995,

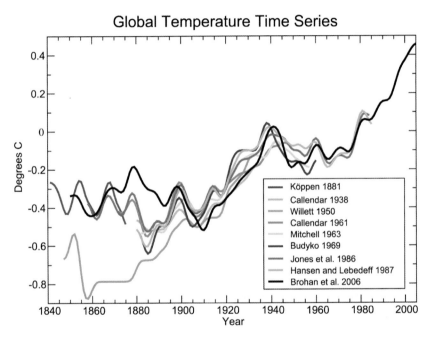

Figure 1. Global temperature, ~1840-2009 AD (IPCC 2007)

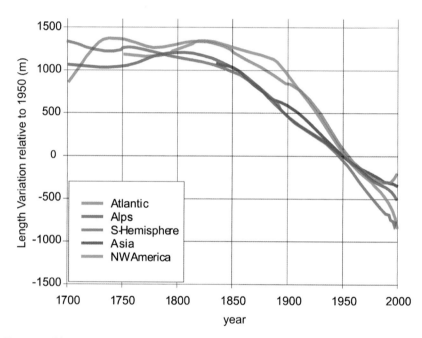

Figure 2. Change in mass of glaciers on land (IPCC 2007)

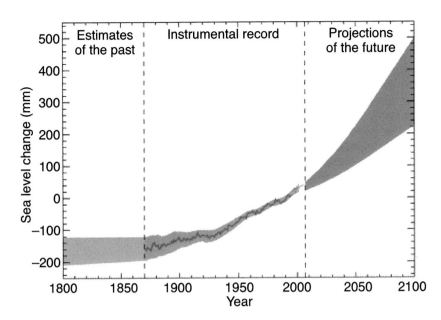

Figure 3. Change in sea level (IPCC 2007)

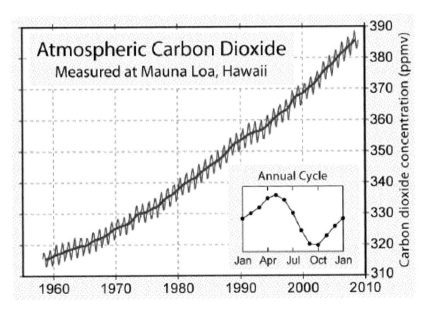

Figure 4. Measured levels of atmospheric CO_2 at the observatory on Mauna Loa, Hawaii (IPCC 2007)

2001, 2007), with progressively greater confidence, that the correlation between rising global temperatures and rising concentrations of CO_2 in the atmosphere should be accorded the status of a direct cause-and-effect relationship, by virtue of (1) the quality of that correlation, and (2) the well-documented physical process whereby large, asymmetrical gas molecules in the atmosphere absorb long-wave (IR) radiation emitted by the solid Earth and delay delivery of that radiation to space at the top of the atmosphere (the "greenhouse effect"). Acknowledging that the progression from correlation to causation represents a conundrum in all branches of science, the IPCC nonetheless invokes a consensus among climate scientists that the primary cause of the contemporary warming trend is the systematic increase of CO_2 (and other heat-trapping gases, primarily CH_4 and N_2O) in the atmosphere, delivered by human activity.

Accordingly, the IPCC has directed an enhanced level of attention toward the construction of multivariate models of atmospheric dynamics and, as the research has advanced, of coupled atmosphere-ocean models that take into account both the thermal inertia of the ocean and the long time scale of oceanic overturn. These models in turn have generated an array of climate conditions that would prevail at specified times in the future, following selected assumptions about the role of "greenhouse gases" compared to other climate processes, and the projected future rate of delivery of those gases to the atmosphere (Fig. 5). All these predictions support the IPCC conclusion that the global climate will grow warmer in the next 100 years (IPCC 1990, 1995, 2001, 2007).

The hazards of extrapolating from the very short instrumental record are well understood. The cooling trend that began about 1941 and persisted until ~1976 (see Fig. 1) encouraged the assertion, briefly discussed in popular media in the mid-1970s, that the warming trend of the previous 130 years had been reversed, and that the world had entered a slow temperature decline (Peterson et al. 2008). That assertion was not widely accepted when it was offered, and was abandoned as temperatures began to rise again in the late 1970s. Indeed, the magnitude of anthropogenic contribution to observed climate change eludes a consensus.

Among those who share the IPCC consensus, the magnitude of the anthropogenic contribution to observed climate change has been difficult to discern. Even among meteorologists committed to instrumental measurements as the only valid descriptor of climate trends, the realization is growing that the one unequivocal statement that can be made about climate is the fact that it changes.

The Record of Sea-Level Change

The temperature measurements that have been made since 1860 (see Fig. 1) are not necessarily representative of the whole Earth, or of the whole atmosphere. A better measure of warming is the rise of global sea level, the consequence of melting of ice on land and return of meltwater to the ocean, and the thermal expansion of surface layers of ocean water in response to warming.

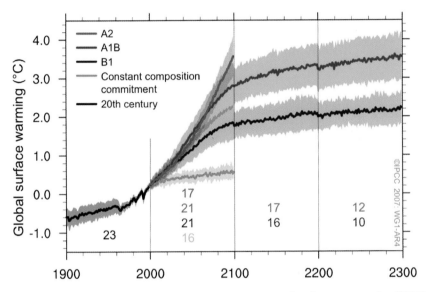

Figure 5. Projections of the contemporary warming trend in future centuries (IPCC 2007). Scenarios A1, A2, and B1 represent three different visualizations of the mix of population growth, economic development, and technological efficiency.

The Constant composition commitment assumes that CO_2 emissions will be reduced to the level that prevailed in 2000 AD.

The black line represents the warming measured through 2000 AD.

Sea level stood ~125 m lower than at present ~18,000 years ago; in the present warming trend, sea level has risen on average 7 mm/yr; the present rate of rise is between 2 and 4 mm/yr, depending on the site where it is measured (Kemp et al. 2009). Even at the highest rate of sea-level rise so far attributed to anthropogenic climate change (Kemp et al. 2009), it will take ~1,500 years for the global sea level to return to the level it occupied 125,000 years ago at the height of the last non-glacial time. Throughout the millennia during which modern human civilization developed, sea level has risen at a rate as fast as, or faster than, the rate at which it is rising today. This realization, however, may offer little consolation today to inhabitants of crowded, low-lying coastal communities.

HISTORY OF UNDERSTANDING OF CLIMATE CHANGE

Awareness that the global climate has been warming in the recent past is not new; abundant historical accounts from the period ca. 1500–1850 in Europe and North America clearly document a time of record cold temperatures that has come to be called the "Little Ice Age" (e.g., Fagan 2001, Lamb 1982). Anecdotal, non-instrumental evidence also clearly documents temperatures warmer than the present for the period of time known as the Medieval Warm Epoch, of ~900–1300 AD (Lamb 1982).

Early in the nineteenth century, Louis Agassiz was made aware that glacial termini in the Alps had retreated noticeably within memory of then-living alpine shepherds; by 1837, Agassiz had learned enough of that phenomenon to assert that the world was then emerging from a period of greater extent of glacier ice on land (e.g., Flint 1971); by 1842, James Forbes had decided to label that earlier period of Earth history the Ice Age (Flint 1971). Further study of the stratigraphic record of glaciation on land, first in Europe and North America and then on many other continents, led to the realization that the Earth has experienced a more or less cyclic history of climate oscillation.

THE DEEP-SEA CORES

The record of oscillation of climate in the recent past is today most fully recorded in the column of off-shore sediment preserved on the sea floor of all the oceans (e.g., Bradley 1999). That sedimentary archive includes mineral organic fragments delivered to the oceans from streams draining the continent; mineral fragments that originated in the water column and similarly accumulated on the sea floor; and the hard parts of plants and animals that carried out their life cycles in the ocean environment and sank to the sea floor upon the death of the organisms that secreted them. The biogenic component of the deep-sea sediment comes primarily from the shallow near-surface layer of ocean water where sunlight penetrates in sufficient magnitude to support photosynthesis and the food chain based on the photosynthate

so produced. Thus, organisms whose remains accumulate on the sea floor have lived more or less in equilibrium with the water column that has supported them, and the chemistry of their shells, and particularly the ratios among oxygen isotopes in those shells, preserves a record of variation in temperature of that water. This record is somewhat compromised by the isotopic composition of glacial meltwaters and the distortion resulting from the isotopic fractionation that occurs as ocean water is evaporated, eventually to make rain and snow.

As these materials accumulate, layer by layer on the ocean floor, they build an archive of ocean-water isotopic history that can be reconstructed by analyzing samples extracted from continuous cores of sea-floor sediment. From that isotopic history a record of ocean-water temperature can be inferred (Fig. 6) and translated into a history of global temperature. Those cores now number in the thousands.

Analysis of the archive provided by those cores has revealed a history of cyclic climatic oscillation that is both longer and more detailed than the record that has been reconstructed from sediments on land, and shows a clear pattern of rise and fall of surface-water temperature of the order of 3° C (Fig. 6). That pattern persists, with variations in amplitude and, to a lesser extent, in periodicity, as far back in Earth history as the cores have been able to

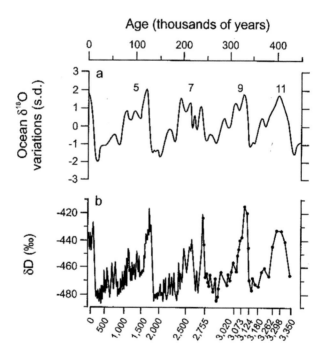

Figure 6. Paleoclimatic information from cores of sea-floor sediment (after Esker 1998).

penetrate, and is apparent in selected suites of sediments from rocks as old as Triassic. That record is now documented from many hundreds of individual cores, and is widely acknowledged to represent both the longest continuous, and the most detailed, archive of past climate history that we have. The chronology of climate variation expressed in those cores has been established by many overlapping geochemical and geophysical geochronometers.

THE CAUSE

From the earliest assertions that global climate is a variable, and that such variation has been preserved in various geologic archives, students of the history of climate have speculated about the probable cause of such variation. As awareness of the cyclic nature of such variation emerged, that speculation turned to a search for processes in nature that might explain the climatic oscillations observed in various geologic archives.

By 1875, Croll had developed a model of the history of climate change based on known geometric variation in the orbit of the Earth around the Sun (Croll 1875; Croll and Irons 1896; Imbrie and Imbrie 1979). Croll's model included both calculations of orbital eccentricity and the role of expanding ice sheets in reinforcing cooling episodes initiated by reduced insolation. His model led him to conclude that glacial periods had been periodic and had been out of phase between the northern and southern hemispheres, and that the last glacial episode had ended 80,000 years ago. The latter conclusion was discredited in 1890 by G. K. Gilbert's celebrated calculation of the post-glacial retreat of Niagara Falls and subsequent radiocarbon determinations of the age of onset of the most recent glacial retreat.

While Croll's work did not stand up to the tests imposed by improved understanding of the timing of climatic variation, his idea that the variation on Earth was driven by subtle variations in delivered insolation controlled by the geometry of the Earth-Sun system rather than by intrinsic variability in solar output was extended and refined by Milutin Milancovič (Milancovič 1941, 1969) in the first half of the twentieth century.

Milancovič undertook to reconstruct the history of intensity of solar radiation striking the Earth, as a function of latitude, season, and geometry of the Earth-Sun orbital system (Milancovič 1941, 1969; Imbrie and Imbrie 1979). Milancovič was aware that the gravitational attractions of Jupiter, Mars, and, to a lesser extent, Venus are sufficient to impose geometric perturbations on Earth's orbit. From what was then known of the orbits of those planets, and the periodic movements of Earth's moon, Milancovič developed a retroactive description of orbital dynamics of the inner solar system over the last 600,000 years. Milancovič proposed that those variations in amount and timing of solar insolation were sufficient to have been the control of the periodicity of glacial oscillations through many climate cycles (Fig. 7); he identified periodicities of 105,000, 41,000, and 21,000 years in that history.

Milancovič's work was first published in 1920 but did not appear in English until 1969 as *Canon of Insolation of the Ice-Age Problem*. Milancovič's work

was completed long before the first measurement was made of $^{18}O/^{16}O$ ratios in foraminiferal shells in deep-sea cores, but was not widely accessed in the USA until after the 1969 translation appeared. By the mid-1970s it was clear, from the close correlation of the Milancovič curves with the emerging data of climatic variation archived in the deep-sea cores, that cycles of intensity of solar insolation (regulated by geometric variation in orbital dimensions, not intrinsic solar luminosity) represented the primary control of climatic oscillation over periods of tens of thousands to hundreds of thousand years. Objections that the magnitude of variation in received solar energy dictated by the orbital variations calculated by Milancovič was insufficient to explain the amplitude of inferred temperature oscillation in the past led to suggestions that various positive feedback processes (see examples below) were capable of amplifying the solar signal.

THE ICE CORES

The record of climate variation preserved in deep-sea cores, as good as it is, has been eclipsed, at least in popular accounts, by the emerging and quite spectacular documentation of the recent history of climate preserved in cores of glacier ice recovered from the Antarctic and Greenland ice sheets (e.g., Petit et al. 1999). The longest of these now spans ~800,000 years, and preserves a detailed temperature record in the isotopic ratios measured in the ice itself, as well as a history of atmospheric composition that can be measured in bubbles of air in the ice.

Each of the ice sheets accumulated as sequential layers of snow, through which superjacent atmosphere circulated freely until the snow was converted to glacier ice by pressure of overlying snow layers, by the process of sublimation/refreezing, and, at least in Greenland, by summer melting and refreezing of meltwater in the following winter. In the Antarctic ice sheet, the snowpack over the ice sheet is converted to glacier ice by those processes operating over a century; thus, the composition of bubbles trapped in the ice represents integrated 100-year averages of atmospheric composition.

The cores from Greenland include enough mineral material to enable glaciologists to identify annual layers of accumulated ice, but the same mineral content reacts chemically with air in the bubbles to confound the efforts to reconstruct atmospheric chemistry; in Antarctica, however, the ice is almost fully devoid of mineral content, and the air bubbles preserve an accurate representation of atmospheric composition. As both the temperature record and the record of atmospheric composition preserved in bubbles trapped in the ice began to emerge through the 1980s, a dramatic correlation was identified: the concentrations of both CO_2 and CH_4 rise and fall with changes in temperature (Fig. 8).

That correlation has been perturbed in the recent past, presumably by the release of both CO_2 and CH_4 by anthropogenic industrial processes. Thus, CO_2, which has varied between 170 ppm and 295 ppm through at least the last eight 100,000-year climate oscillations, was measured empirically in the atmosphere at 315 ppm in 1958, and now stands at 385 ppm (see Fig. 4); CH_4,

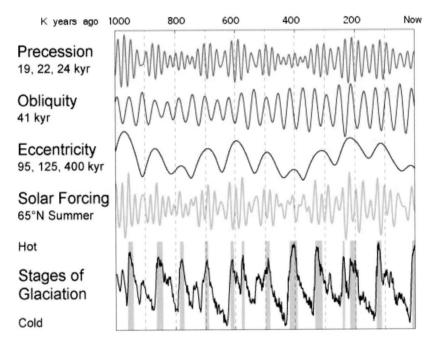

Figure 7. The Milankovič cycles (Milankovič 1941, 1969).

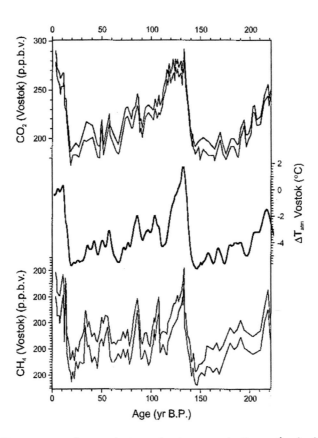

Figure 8. Temperature data and atmospheric concentrations of principal greenhouse gases from analysis of the Antarctic ice cores (IPCC 2007).

which has varied from 400 ppb to 700 ppb through eight full cycles in the ice cores, now stands at 1750 ppb (IPCC 2007).

While the recent anomalous concentrations of both gases can be attributed to anthropogenic processes, those sources cannot be responsible for the lock-step correlation between temperature of the ice and concentration of the gases CO_2 and CH_4 in that ice through many climate cycles, long before humans began to extract and use fossil fuels. That pre-industrial correlation among these variables has been explained through several positive feedback loops.

1. The solubility of gases is inversely proportional to temperature; thus, in times when the ocean is warming, CO_2 and other gases are driven out of solution in the ocean into the atmosphere; at times when the ocean is cooling, those gases sink into the ocean. That process has been quantified by measuring the migration into the water column of both ^{14}C-depleted CO_2 from the combustion of fossil fuel in which effectively all ^{14}C has decayed (the Suess Effect; see Keeling, 1980), and ^{14}C-enriched CO_2 from the

excess inventory of ^{14}C produced by neutrons emitted during atmospheric testing of nuclear devices between 1945 and 1963 (Broecker 1998, 2005).

2. Breakdown of leaf litter by soil micro-organisms is enhanced by higher temperatures more than photosynthetic drawdown of CO_2; thus, in warm times CO_2 is driven out of soils into the atmosphere, but in cool times photosynthetic drawdown of CO_2 predominates.

3. When soils saturated with water freeze, gas exchange with the atmosphere is effectively terminated; decay of organic matter continues slowly, however, and becomes fermentation when soil supplies of O_2 are exhausted. Thus, no CO_2 escapes to the atmosphere from frozen saturated soils, but C is delivered to the atmosphere as CH_4 when those soils thaw in warming times. The CH_4 so released is converted to CO_2 and H_2O, with a mean residence time of ~7 years.

Two pieces of crucial information emerge from consideration of the data from the ice cores.

1. Atmospheric concentrations of both CO_2 and CH_4 are higher today than at any time represented in the 800,000 years of ice-core records.

2. Long before human activity began transferring carbon from reserves of fossil fuel into the atmosphere, natural processes were moving carbon among Earth-surface reservoirs in large volumes, and the sizes of the carbon inventories in those reservoirs changed in response to changes in temperature.

Given that the fundamental control of climate through the period represented in the ice cores is widely recognized to be the rhythm imposed by the Croll-Milancovič cycles, we must conclude from the pre-industrial record in ice cores that global temperature has governed the atmospheric concentration of CO_2, not the reverse.

The Carbon Cycle

The mass of carbon in Earth-surface reservoirs (lithosphere, hydrosphere, atmosphere, biosphere) is not large. Carbon moves among Earth-surface reservoirs by many processes and at variable rates. Figure 9 is a box model of the carbon cycle, on which current estimates of both masses of carbon in Earth-surface reservoirs and annual fluxes between and among those reservoirs are shown. The mass of carbon in the atmosphere is increasing at the rate of ~4.1 Gt/yr.

Detailed calculations of the mass of carbon delivered to the atmosphere each year by anthropogenic processes have been carried out since Charles Keeling began making measurements of atmospheric concentrations of CO_2, first at an observatory on Mauna Loa, HI, and the South Pole, and later at many locations scattered across the globe (Keeling et al. 1976; Keeling and Whorf 2005). Those measurements, and Keeling's calculations of the volumes of fossil fuels extracted and burned by human industry since the start of the

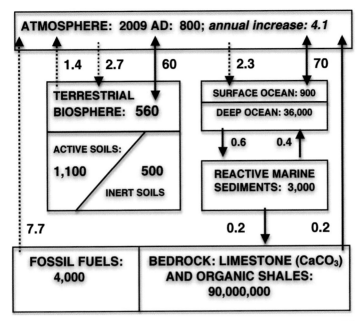

Figure 9. A box model of the carbon cycle, compiled from many sources. The mass of carbon (C) in each reservoir is expressed as gigatonnes (Gt) C. Fluxes between reservoirs (arrows) are expressed in Gt C/yr. The anthropogenic fluxes (dotted arrows), and the annual increase in atmospheric C, represent the best estimates for 2000–2008 by the Global Carbon Budget Consortium (2009).

industrial revolution, reveal a 200-year history of perturbation of fluxes in the carbon cycle. Those calculations also show that the human contribution is small, compared to so-called natural fluxes.

The question of whether or not, and if so, to what extent, the excess CO_2 delivered to the atmosphere by industrial processes has increased the mean surface temperature of Earth through an enhanced greenhouse effect is not always seen as worthy of reasoned debate. Yet that question is not as clear cut as advocates in the two extreme positions in the current debate would argue. That the anthropogenic contribution is measurable is apparent from one consideration:

During each of the previous deglacial episodes documented in the ice cores, CO_2 and CH_4 were driven into the atmospheric reservoir from other Earth-surface reservoirs, following the feedback processes described above. During the last ~100 years of the present deglacial period, that net flux has been reversed: the "excess" CO_2 is sinking into one or another reservoir *against* the thermally driven gradient that controlled flow of these gases in previous deglacial times. This reversal of flux represents the strongest evidence that human industrial activity has altered processes in the climate system.

Several research groups have undertaken to reconstruct the history of the CO_2 concentration in the atmosphere from various proxy archives: (1) the

density of leaf stomata (stomatal area/total leaf area) in fossil leaves, which, from analogy with living leaves, represents a tradeoff between water loss and photosynthetic efficiency (e.g., Royer et al. 2001, 2007); and (2) complex models of partition of C among Earth-surface reservoirs, based on stratigraphic balances and ratios among carbon isotopes in fossil carbonates (Berner and Kothavala 2001). These exercises have led to a reconstruction of the history of CO_2 in the atmosphere (Fig. 10) that shows that levels of atmospheric CO_2 through the period of time represented in the ice cores are among the lowest recorded in Phanerozoic time.

This inference is further supported by the fact that isotopic evidence of C_4 photosynthesis, acknowledged by botanists to represent an adaptive response to diminishing levels of CO_2 in the atmosphere, appears in the fossil record only 8 million years ago. The first appearance of C_4 photosynthesis is interpreted to represent a response of the biosphere to declining levels of CO_2 through the latter part of Tertiary time (Ehleringer et al. 1997).

At least one provocative hypothesis addresses directly the extent of human perturbation of atmospheric chemistry in prehistory: the suggestion by Ruddiman (2003, 2007) that humans began to alter the chemistry of the atmosphere when they invented agriculture, which delivered unprecedented levels of CH_4 to the atmosphere from wet-rice culture in southeast Asia as early as 7,000 BP, and added measurable amounts of CO_2 to the atmosphere, the consequence of land clearing and soil tillage, as early as 5,000 years BP. Ruddiman

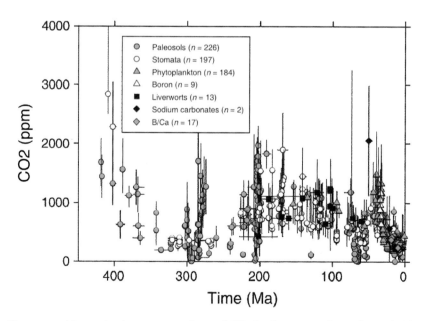

Figure 10. Atmospheric concentrations of CO_2 in the atmosphere through Phanerozoic time (updated from Royer et al. 2007).

has identified short-term reductions in anthropogenic output of CO_2 and CH_4 corresponding to documented periods of sharp reduction in human population, such as those that occurred in the several European plagues and in the depopulation of the Americas after European contact.

Ruddiman's hypothesis has not received the attention its implications merit. Many recent studies, notably those summarized by the IPCC (1990, 1995, 2001, 2007), emphasize the human impact on atmospheric chemistry since the start of the Industrial Revolution and the implications of that escalating phenomenon for future human generations, and fail to consider that a far smaller, technologically undeveloped human population might have altered the chemistry of the atmosphere long ago. The community of scientists skeptical of the assertion that human society has altered atmospheric chemistry is even less likely to be persuaded that a far smaller human population has had a measurable effect on atmospheric chemistry so long ago.

In the present volume, we consider the effect on human communities of climate change that can be documented from archeological and geological archives. Ruddiman's assertion adds another dimension to the discussion by presenting evidence that the land-use changes that accompanied implementation of agriculture imposed a quantifiable effect on atmospheric chemistry (2003, 2007).

Other Factors

While the dramatic correlation among isotopic ratios in deep-sea cores, Milancovič cycles, and temperature data in the ice cores has led students of climate to express some confidence that the mechanisms that control climatic variation are understood on timescales of tens of thousands to hundreds of thousands of years, less is known about the influence on climate of factors that, if cyclic, operate over periods of time shorter than the solar-orbital cycles of Milancovič (1941, 1969).

Foremost among these factors is the role of the oceans in storing solar heat at locations of high insolation, transporting that heat across global trajectories via both surface and thermohaline circulation processes, and transferring heat back to the atmosphere at locations that receive lower levels of insolation. Far more efficiently than the atmosphere, the ocean, by virtue of the greater thermal inertia of water compared to air, and the great thermal mass represented by the ocean, regulates Earth-surface temperature by moving equatorial heat toward the poles on a scale that dwarfs the heat-transfer capacity of the atmosphere.

Oceanographers understood the geometry of wind-driven surface currents in the ocean long before they had begun to chart the density-driven deeper circulation vectors that mix solar energy through the water column, delivering equatorial heat to higher latitudes on a time scale of ~1,500 years. As understanding of ocean circulation has grown, attention has been drawn to a number of short-term oscillations that appear to be at least semi-periodic over periods of 5–30 years, superimposed on this larger circulation pattern.

These oscillations include the El Niño–Southern Oscillation (ENSO) phenomenon and the North Atlantic Oscillation (NAO), neither of which has yet yielded a dependable periodicity (Hurrell et al. 2005), and the Pacific Decadal Oscillation (PDO), with the suggestion of a decadal or multi-decadal periodicity (Trenberth and Hurrell 1990; Bond and Harrison 2000).

The role of these short-term oceanic variables in the control of climate is not yet well known, but they appear not to have been important enough to have overprinted the correlation among the temperature records in the ice cores, the deep-sea cores, and the extrapolated temperature from the Milancovič calculations.

Many other factors have been shown to exert an influence on circulation of the atmosphere and distribution of solar heat across various provinces. Numerous historic observations have shown that documented individual explosive eruptions of subduction-zone volcanoes have thrown enough particulate matter into the atmosphere to influence climate, both locally and, in a few cases, globally, for a matter of years. While these events have been dramatic and, because of their capacity to disrupt the normal routines of human affairs, have been documented in detail, there is no evidence that volcanicity has followed temporal patterns in the past sufficient to impose a control on documented climate variation.

The model of plate tectonics, in which volcanic eruptions occur at spreading zones, at subduction zones, and at intraplate hot spots, requires that, over any extended period of time, the amount of volcanic rock produced at spreading zones will approximately equal the amount of rock lost to the mantle at subduction zones; as long as the rate of sea-floor spreading is a constant, which it has approximated for the last 180 million years, the model does not allow much variation in the production rates of volcanic rock, and hence not much variation in the amount of volcanic rock extruded per unit time.

Other terrestrial factors include delivery of dust to the atmosphere, and the rise of mountain ranges with the capacity both to alter patterns of circulation of the atmosphere and to present fresh rock to weathering processes, with consequent complexation of atmospheric CO_2 as carbonate weathering products are formed. The dust that we now see swept from the Sahara by easterly trade winds and carried around the world to serve as condensation nuclei for rainfall at distant locations is probably linked to one or another feedback loop in the climate system. The rise of mountain ranges, certainly important in controlling climate over long periods of Earth history, is too slow to have been an important factor in the time span represented in our primary archives (deep-sea cores and ice cores).

Variation in output of the Sun has been discussed since the earliest years of speculation as to the causes of climatic variation, but has not been measured with precision sufficient to separate signal from noise until the advent of orbiting satellites with the capacity to measure insolation directly above the confounding influence of a dynamic atmosphere that is variable in both depth and composition.

The best known of cyclic variations of solar output is the variation in the number of sunspots, now well documented for at least 400 years (e.g., Schröder

2005; Dean 2005; Ochadlick et al. 1993; Hughes et al. 2003), with clear period-icities of ~11 years, ~70–90 years, and ~210 years. At least five periods of very low sunspot numbers have been documented; among them, the Maunder Minimum (1645–1715) has been proposed as a driving factor for the coldest time within the Little Ice Age. It seems likely that these short-term variations in solar activity are superimposed on longer cycles not yet well defined, given the short period of direct human observation of solar phenomena (Vita-Finzi 2008 see also Manning, Ch. 2, this volume).

While it does seem that variation of solar output, on many scales, must have had an effect on global climate, those factors, like other short-term variables discussed here, have not overprinted the pattern in the three principal archives.

Summary

The climate of Earth is not stable. It has not been stable within the record of climate history preserved in various geologic archives.

The globe is warming today, as demonstrated by many lines of evidence. Among those lines of evidence the best integrated expression of global warming is the rate of rise of sea level.

The present period of rising temperature and rising sea level began ~18,000 years ago.

For at least the last 800,000 years, and very likely for a much longer time, Earth has experienced a sequence of periodic changes in absorbed insolation that have driven a series of high-amplitude climatic oscillations with principal periodicities of ~105,000, 41,000, and 21,000 years.

The amplitude of this oscillation has been reinforced by several positive feedback processes, one of which has been the release of greenhouse gases (CO_2, CH_4) in periods of rising temperature. The increase of the concentrations of greenhouse gases in the atmosphere has probably further reinforced the warming at those times; the decline in concentrations of those gases during cooling times has probably reinforced the cooling trend. The extent of that reinforcement has not been established.

For at least the last 800,000 years, and very likely for a much longer part of Cenozoic time, the temperature regime on Earth has remained cold enough to support permanent ice sheets at both poles. The oscillation in insolation at those times found expression in the advance of continental glaciers in cold periods and the wholesale retreat of those glaciers in warmer times, but the ice at the poles did not fully melt at any time in the last 800,000 years.

The circumstances of permanent ice sheets at the poles, well documented for late Cenozoic, late Paleozoic, and Ordovician time, have been met for no more than 15% of accessible Earth history.

Superimposed on the fundamental climatic oscillations has been a series of climate changes that are less well understood and that, if cyclic, follow much shorter periods. Those climatic oscillations cannot be attributed to the variations in orbital geometry identified by Milancovič and may be controlled by forcing factors that are not obviously periodic.

In the last 200 years, the accelerated development of human material culture has altered the composition of the atmosphere to an extent unprecedented in the 800,000-year record in the ice cores. It is reasonable to draw the inference that anthropogenic release of CO_2 from the burning of fossil fuels over the last 200 years is responsible for the observed increase in the atmospheric inventory of CO_2. It is also reasonable to infer that the observed warming trend of the last ~50 years is at least partly a direct consequence of an enhanced greenhouse effect. It is clearly prudent to reduce, to the extent possible, the single potential driver of climate change over which we can exercise some control, while we continue efforts to understand the dynamics of the climate system.

DISCUSSION

We know that the climate of Earth is changing. We know that it has always changed. The processes and factors that we now recognize to exert some control over the climate change now under way have controlled that process throughout human history, and for many millions of years before the emergence of our hominin ancestors. Indeed, the adaptive response of our prehuman ancestors to the variation in physical conditions in the environment has made us what we are and determined where and how we live.

Many governments and private institutions have invested heavily in assessing the risk of future climate change, and in developing plans to mitigate the anthropogenic contribution to that change. Less has been invested in preparing human society for the climate change that looms inevitably in our future. Scientists and political leaders speak of stabilizing climate; do they realize that climate has never been stable? Ambitious futurists speak of engineering future climate to our specifications; do they realize that our record of engineering the natural environment to our own liking has very often led to unintended adverse consequences?

Moreover, while the attention of scientists and planners everywhere has been directed at the need to prepare human society for climate crises of the near future, we have not directed as much attention to the history of human response and adaptation to the changes in climate that we know have occurred repeatedly in the past. Those responses are recorded throughout written human history, and for earlier times can be reconstructed, with more or less accuracy, from the archive of past environments preserved in archeological and geological records.

Conclusions drawn from such empirical evidence may provide a useful corrective to computer models of future climate based on short-term climatic projections.

REFERENCES

Berner, R. A., and Z. Kothavala. 2001. GEOCARB III: a revised model of atmospheric CO_2 over Phanerozoic time. *American Journal of Science* 301: 182–204.

Bond, N. A., and D. E. Harrison. 2000. The Pacific decadal oscillation, air-sea interactions and central north Pacific atmospheric regimes. *Geophysical Research Letters* 27: 731–34.

Bradley, R. S. 1999. *Paleoclimatology: reconstructing climates of the Quaternary.* San Diego: Academic Press.

Broecker, W. S. 1998. *The greenhouse puzzle.* New York: El Digio Press.

———. 2005. *The role of the ocean in climate: yesterday, today, and tomorrow.* New York: El Digio Press.

Croll, J. 1875. *Climate and time, in their geological relation: a theory of secular changes of the earth's climate.* London: Daldy, Isbister & Co.

Croll, J., and J. C. Irons, 1896. *Autobiographical sketch of James Croll: with a memoir of his life and work.* London: Edward Stanford.

Dean, W. E. 2005. The sun and climate. *U.S. Geological Survey Fact Sheet* 0095-00.

Ehleringer, J. R., T. E. Cerling, and B. R. Helliker. 1997. C4 photosynthesis, atmospheric CO_2, and climate. *Oecologia* 112: 285–99.

Esker, D., G. P. Eberli, and D. McNeill. 1998. The structural and sedimentological controls on the reoccupation of Quaternary incised valleys, Belize Southern Lagoon. *Bulletin of the American Association of Petroleum Geologists* 82: 2075–2109.

Fagan, B. M. 2001. *The little ice age: how climate made history, 1300–1850.* New York: Basic Books.

Flint, R. F. 1971. *Glacial and Quaternary geology.* New York: John Wiley & Sons.

Global Carbon Budget Consortium. 2009. Carbon 2008 Budget: http://www.globalcarbonproject.org/carbonneutral/index.htm.

Hughes, G. B., R. Giegengack, and H. N. Kritikos. 2003. Modern spectral climate patterns in rhythmically deposited argillites of the Gowganda Formation (early Proterozoic), southern Ontario, Canada. *Earth & Planetary Science Letters* 207: 13–22.

Hurrell, J. W., Y. Kushnir, G. Ottersen, and M. Visbeck. 2005. An overview of the North Atlantic Oscillation. In *The North Atlantic Oscillation: climatic significance and environmental impact,* ed J. W. Hurrell, Y. Kushnir, G. Ottersen, and M. Visbeck, 1–35. Geophysical Monograph 134, Washington: American Geophysical Union.

Imbrie, J., and K. Imbrie. 1979. *Ice ages: solving the mystery.* Cambridge, Mass.: Harvard University Press.

IPCC (Intergovernmental Panel on Climate Change). 1990, 1995, 2001, 2007, 2010. *Climate change.* Cambridge: Cambridge University Press.

Jansen, E., J. Overpeck, K. R. Briffa, J.-C. Duplessy, F. Joos, V. Masson-Delmotte, D. Olago, B. Otto-Bliesner, W. R. Peltier, S. Rahmstorf, R. Ramesh, D. Raynaud, D. Rind, O. Solomina, R. Villalba, and D. Zhang. 2007. Palaeoclimate. In *Climate change 2007: the physical science basis: contribution of working group I to the Fourth Assessment Report of the Intergovernmental Panel on Climate Change,* ed. S. D. Qin, M. Manning, Z. Chen, M. Marquis, K. B. Avery, M. Tignor, and H. L. Miller, 433–97. Cambridge: Cambridge University Press.

Keeling, C. D. 1980. The Suess effect: [13]carbon-[14]carbon interrelations. *Environment International* 2: 229–30.

Keeling, C. D., R. B. Barcastow, and T. P. Whorf. 1976. Atmospheric carbon dioxide variations at Mauna Loa Observatory. *Tellus* 28: 538–51.

Keeling, C. D., and T. P. Whorf. 2005. Atmospheric CO_2 records from sites in the SIO air sampling network. In *Trends: a compendium of data on global change*. Oak Ridge, Tenn.: U.S. Department of Energy. http://cdiac.esd.ornl.gov/trends/co2/sio-keel-fl ask/sio-keel-fl ask.html.

Kemp, A. C., B. P. Horton, S. J. Culver, D. R. Corbett, O. van de Plassche, W. R. Gehrels, and B. C. Douglas. 2009. The timing and magnitude of accelerated sea-level rise. *Geology* 37: 1035–38.

Lamb, H. H. 1982. *Climate, history and the modern world*. London: Routledge.

Milancovič, M. 1941. *Kanon der Erdbestrahlung und seine Anwendung auf das Eiszeitenproblem* (translated into English, 1969, as *Canon of insolation of the ice-age problem*). Washington, D.C.: Israel Program for Scientific Translations, U.S. Department of Commerce and the National Science Foundation.

Ochadlick, A. R., Jr., H. N. Kritikos, and R. Giegengack. 1993.Variations in the period of the sunspot cycle. *Geophysical Research Letters* 20: 1471–74.

Peterson, T. C., W. M. Connolly, and J. Fleck. 2008. The myth of the 1970s cooling scientific consensus. *Bulletin of the American Meteorological Society* 89: 1325–37.

Petit, J. R., J. Jouzel, D. Raynaud, N. I. Barkov, J.-M. Barnola, I. Basile, M. Bender, J. Chappellaz, M. Davis, G. Delaygue, M. Delmotte, V. M. Kotlyakov, V. Lipenkov, C. Lorius, L. Pepin, C. Ritz, E. Saltzman, and M. Stievenard. 1999. Climate and atmospheric history of the last 420,000 years from the Vostok Ice Core, Antarctica. *Nature* 399: 429–36.

Royer, D. L., R. A. Berner, and D. J. Beerling. 2001. Phanerozoic atmospheric CO_2 change: evaluating geochemical and paleobiological approaches. *Earth Science Reviews* 54: 349–92.

Royer, D. L., R. A. Berner, and J. Park. 2007. Climate sensitivity constrained by CO_2 concentrations over the past 420 million years. *Nature* 446: 530–32.

Ruddiman, W. F. 2003. The anthropogenic greenhouse era began thousands of years ago. *Climate Change* 6: 261–93.

———. 2007. *Plows, plagues, and petroleum: how humans took control of climate*. Princeton: Princeton University Press.

Schröder, W. 2005. *Case studies on the Spörer, Maunder, and Dalton minima*. Potsdam: Beiträge zur Geschichte der Geophysik und Kosmischen Physik 6.

Thompson, L. G., E. Mosley-Thompson, H. Brecher, M. E. Davis, B. Leon, D. Les, T. A. Mashiotta, P.-N. Lin, and K. Mountain. 2006. Evidence of abrupt tropical climate change: past and present. *Proceedings of the National Academy of Sciences* 103: 10536–543.

Trenberth, K. E., and J. W. Hurrell. 1994. Decadal atmosphere-ocean variations in the Pacific. *Climate Dynamics* 9: 303–19.

Vita-Finzi, C. 2008. *The sun: a user's manual*. New York: Springer.

2

Radiocarbon Dating and Climate Change

STURT W. MANNING

INTRODUCTION

Radiocarbon is the most widely used dating technique—and, hence, fundamental chronometric framework—for the Holocene period in archaeology and for many environmental sciences.[1] But radiocarbon is not a neutral chronicler of time. Instead, it offers a record of past cosmic ray interaction with the Earth's atmosphere, modulated over the shorter term by changes in solar magnetic activity. Hence, we have a wiggly radiocarbon calibration curve, and the irregular transfer of radiocarbon ages to calendar ages. The consequence, if we look back in time armed with radiocarbon, is that not all time is equal. In which case, if we are usefully to consider past climate crises or climate changes (whether good or bad for a given society or region) or other key archaeological markers in a wider perspective, as discussed in this volume, we should consider the chronometric viewpoint (or lens) provided by radiocarbon when we try to date specific episodes in past time, or when we assess the dated information we have on offer.

THE RADIOCARBON RECORD[2]

Naturally occurring radiocarbon (^{14}C) is a radioisotope produced when cosmic rays enter the Earth's atmosphere. It was originally assumed that production was constant, and that the ^{14}C activity of the atmosphere was in equilibrium with the biosphere and oceans. However, we now know that production varies over time. Modulation of the cosmic ray flux due to changes in solar magnetic activity is the main shorter-term variable during the last 10,000 years. Over longer-term timescales the Earth's geomagnetic field is seen as the key controlling regime. Thus, for the Holocene especially, the radiocarbon record

*I thank Bernd Kromer for collaboration and discussion on the intersection of radiocarbon and tree rings, for providing various datasets, and for comments on an earlier version of this text. I thank Mary Jaye Bruce for improving the text.

25

is not only critical for dating; it is also a proxy for variations in solar activity. From observations of the last few centuries, shorter-term cycles or variations in solar activity are known, and from ^{14}C and ^{10}Be longer-term cycles are observed: including an apparent recurrent cycle of around 1500 years that seems to be associated with key Holocene climate change episodes—a topic to which we will return below.

The other major variable in the observed radiocarbon record is the oceans. On the Earth, since the vast majority of ^{14}C (i.e., $^{14}CO_2$) ends up in the oceans, changes in exchange of CO_2 between the oceans and the atmosphere will also affect ^{14}C levels in the atmosphere, and here, too, research subsequent to the original conception of the ^{14}C method has shown that important variations occur. The sharp rise in ^{14}C levels in the atmosphere associated with the onset of the cold period known as the Younger Dryas has been attributed mainly to a shutdown (or substantial reduction) in ocean overturning circulation and linked changes (Hughen et al. 2000; Delaygue et al. 2003). At a more modest scale, patterns evident in the radiocarbon record following major cooling or warming episodes imply ocean modulation (as do the changing offsets—smaller during cooler periods, larger during warmer periods—between the Northern and Southern Hemisphere records; see Figs. 1–2).

For the study of the terrestrial world in the Holocene, we are interested in the record of atmospheric ^{14}C. The high-resolution archive for investigation of this record comes from analysis of time-series of known-age tree rings. In this chapter, I only consider the mid-latitudes of the Northern Hemisphere, where there is a recognized record of atmospheric radiocarbon based on measurements on wood at several laboratories over the last several decades back to the eleventh millennium BC (Reimer et al. 2004—the IntCal curve is periodically updated to reflect additional and improved data and analysis; a Southern Hemisphere curve exists for the recent millennia; see Figs. 1–2). This radiocarbon calibration curve provides best estimates for radiocarbon levels in the atmosphere back through time, and, since these have varied in line with changing solar activity and other factors, it yields a wiggly graph (Figs. 1–2).

This radiocarbon calibration curve is our chronological reality. Any given radiocarbon age on a dated sample must be calibrated against this calibration curve to determine the range of calendar years to which the sample belongs (and the probabilities belonging to the various years within that range). The wiggly shape of the calibration curve (that is, the past history of atmospheric radiocarbon levels) means that this is a non-monotonic function, and that there are widely varying possible outcomes. Three differing examples of radiocarbon date calibration are shown in Figures 3–5.

The shape of the radiocarbon calibration curve therefore makes dating relatively precise in some periods, and less precise to difficult or ambiguous in others (Guilderson et al. 2005; Weninger 1990). This "boon and bane of radiocarbon dating" (Guilderson et al. 2005) can be quantified and illustrated (see also Guilderson et al. 2005, 363, upper figure; Figs. 6 and 7). We see that not all time is equal (meaning that radiocarbon-based time has preferences and non-preferences) in terms of dating (and so can even be described as biased;

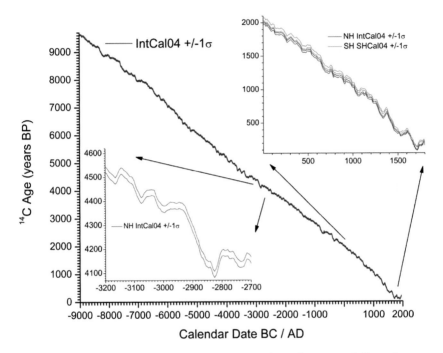

Figure 1. The Northern Hemisphere IntCal04 radiocarbon record (from known-age tree rings) in blue shown as a 1 Standard Deviation (1σ) curve over the last 11,000 years (Reimer et al. 2004). The insets show the curve in more detail over two shorter time intervals (with the Southern Hemisphere calibration curve also shown in red for the period AD 1–1800; McCormac et al. 2004); the period 3200–2700 BC (bottom left) is chosen as modern high-precision measurements of the marked variations across this period (as part of the period 3200–700 BC) formed an important part in confirming the truly wiggly (non-monotonic) nature of the radiocarbon record (de Jong et al. 1979).

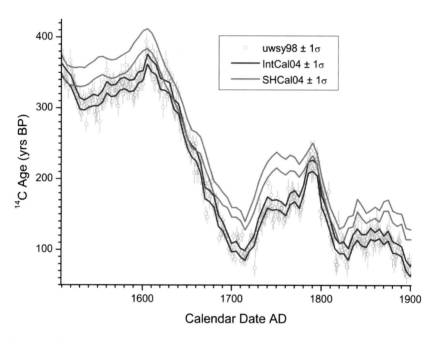

Figure 2. Radiocarbon calibration data for the period AD 1500–1900, comprising an annual record (Stuiver et al. 1998a), the Northern Hemisphere IntCal04 record (Reimer et al. 2004), and the Southern Hemisphere record (McCormac et al. 2004). Note how the contemporary difference (vertical distance) varies between the Northern and Southern Hemisphere datasets (see text).

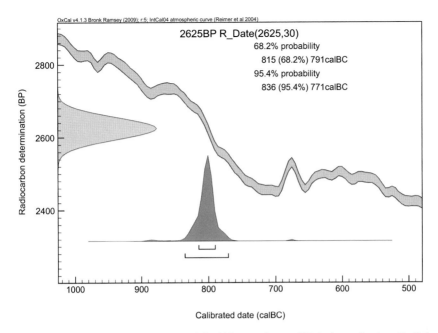

Figure 3. Calendar age calibration of the ¹⁴C age 2625±30 BP (1σ) employing OxCal (Bronk Ramsey 1995, 2009) and IntCal04 (Reimer et al. 2004). Note how the steep slope in the radiocarbon calibration curve at this period concentrates the ¹⁴C age range into a narrower and very specific calendar age range (1σ calendar range of 24 years versus a 60 ¹⁴C years range at 1σ). Calibrated ages are expressed either in cal years BC/AD or Cal years BP (Before Present = AD 1950 for radiocarbon).

Walanus 2009), whether from the perspective of the radiocarbon timescale (Fig. 6) or from the perspective of the calendar timescale (Fig. 7).

Inspection of Figures 6 and 7 highlights that some specific calendar periods are much more likely to be dated—in particular, those periods lying on the steep slopes in the radiocarbon calibration curve, such as those shown either side of 2900 BC in Figure 1 Lower Inset, AD 1650 in Figure 2, or 800 BC in Figure 3. Meanwhile, some other periods are less likely to be dated. This uneven distribution of dating probability will, in turn, promote a tendency for the likely outcome periods to suck-in and claim associated events, and for these episodes to become more important than they necessarily should be (and the reverse).

Hence, where major climate change episodes correlate with major steep slopes in the radiocarbon calibration curve, we should be able to find the episodes and indeed may well exaggerate their prominence. And, conversely, where major climate events coincide with wiggles and inversions in the radiocarbon calibration curve (perhaps even because the event itself causes the inversion, as when a major input of subglacial meltwater entered the North Atlantic c. 8200 BP; Daley et al. 2009), then it will be more difficult to

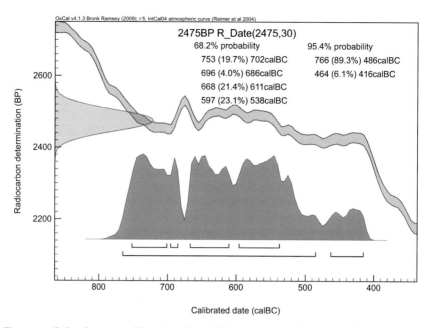

Figure 4. Calendar age calibration of the ^{14}C age 2475±30 BP (1σ) employing OxCal (Bronk Ramsey 1995, 2009) and IntCal04 (Reimer et al. 2004). Note how the plateau in the radiocarbon calibration curve at this period spreads the ^{14}C age range into a much wider and less specific calendar age range (1σ range of 60 ^{14}C years, versus 1σ range covering 213 calendar years).

resolve these events. The much discussed episode of abrupt climate change dating about 6250 BC (or 8200 CalBP) may be a case in point.[3] If we acquired a good radiocarbon date on a short-lived sample (meeting all quality control expectations; Boaretto 2009) corresponding to this event (with a σ of ±25 ^{14}C years), the shape of the radiocarbon calibration curve at this period would spread the corrected age into two regions (Fig. 8). And if we consider a range of real calendar ages for this episode from 8250 to 8150 CalBP, the resolved range on the same basis will spread over three centuries to span the period c. 8336–8022 CalBP at 2σ (8319–8035 CalBP at 1σ). This makes it difficult to date events falling within the 8200 CalBP interval without additional information (high-resolution data, extra parameters, and detailed analysis; e.g., Blaauw et al. 2007; Blaauw and Christen 2005). And, even then, it still may prove challenging to overcome the basic underlying dating ambiguity fully. For example, let us hypothesize a stratigraphic sequence of 9 units from A, the oldest, to I, the youngest, with unit E corresponding to 8200 CalBP and covering in total around 350 or so calendar years and 400 ^{14}C years in total. Two good modern AMS radiocarbon dates are run on ideal short-lived samples for each unit (eighteen dates in all) to yield weighted average ages for each unit at ±18 ^{14}C years BP. We find that we obtain the hypothetical and overly perfect sequence of:

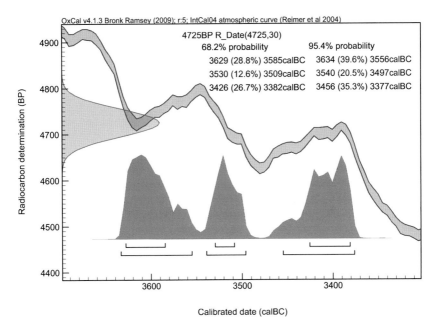

Figure 5. Calendar age calibration of the ¹⁴C age 4725±30 BP (1σ) employing OxCal (Bronk Ramsey 1995, 2009) and IntCal04 (Reimer et al. 2004). Note how the wiggles in the radiocarbon calibration curve at this period spread and divide the original single ¹⁴C age range into three different possible calendar age ranges covering overall a much wider and less specific total calendar age range.

Unit A:	7615 ± 18 ¹⁴C years BP
Unit B:	7565 ± 18 ¹⁴C years BP
Unit C:	7515 ± 18 ¹⁴C years BP
Unit D:	7465 ± 18 ¹⁴C years BP
Unit E:	7415 ± 18 ¹⁴C years BP
Unit F:	7365 ± 18 ¹⁴C years BP
Unit G:	7315 ± 18 ¹⁴C years BP
Unit H:	7265 ± 18 ¹⁴C years BP
Unit I:	7215 ± 18 ¹⁴C years BP

If this stratigraphic sequence of data is then analyzed employing Bayesian methods to try to best resolve unit E, the modelled calendar age range for unit E is 8306–8184 BC at 2σ (8296–8187 BC at 1σ) (Fig. 9), only slightly better than the raw range found in Figure 8 and failing again specifically to resolve 8200 CalBP.

This issue is in fact a very real world problem. Consider the identification by Daley et al. (2009) of a terrestrial signal for the 8200 year CalBP cold event. These authors had twenty-three radiocarbon dates from a core, NDN02/1 (compare Fig. 9). Their (non-Bayesian, curve fitting) age-depth model placed

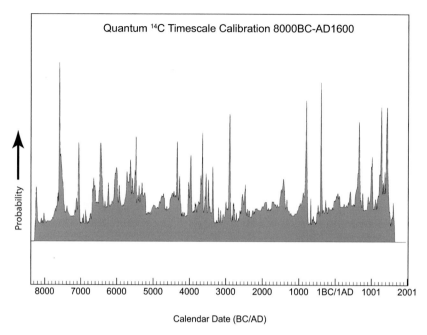

Figure 6. Combined calendar dating probability histogram for radiocarbon dates at approximately maximum current AMS precision (±25 ^{14}C yrs BP) evenly spaced every 5 ^{14}C years across the radiocarbon timescale covering the period 8000 BC to AD 1600, using OxCal (Bronk Ramsey 1995, 2009) and IntCal04 (Reimer et al. 2004) with calibration curve resolution set at 5.

the major δ^{18}O precipitation event that they linked with the 8200 CalBP event (640–647cm depth) at about 8338–8391 CalBP. Daley et al. (2009, 834) thus conclude that "it remains to be explained why the signal of the 8200 yr B.P. event in Greenland ice cores occurred ~200 yr later than in NDN02/." This seems an important and challenging finding—and the situation seems to call for consideration of mechanisms that could cause such an offset. However, the specifically relevant radiocarbon evidence on either side of the 640–647cm depth section of the core in fact yielded ^{14}C dates that could reduce this apparent two-hundred-year offset dramatically almost to zero (see Fig. 10). The calibration curve properties combined with the curve-fitting exercise in Daley et al. (2009) favored the higher possible ages. This could thus be a ^{14}C bias issue—and not necessarily a key offset in data records that needs some other explanation.

To try to overcome this problem, we can consider a comprehensive Bayesian analysis of an age-depth model (Bronk Ramsey 2008)—instead of the ad hoc curve fitting to the mid-points of the calibrated ranges as employed by Daley et al. (2009)—employing the data in the Daley et al. (2009) paper (Fig. 11). The analysis can also be used to calculate the calibrated age ranges to be interpolated for depths without radiocarbon data, that is depths 640cm and 647cm

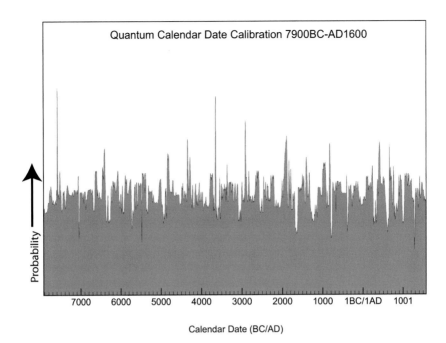

Figure 7. Combined calendar dating probability histogram for simulated radio-carbon dates at approximately maximum current AMS precision (±25yrs ^{14}C BP) evenly spaced every 5 calendar years across the radiocarbon timescale covering the period 7900 BC to AD 1600, using OxCal (Bronk Ramsey 1995, 2009) and Int-Cal04 (Reimer et al. 2004) with calibration curve resolution set at 5.

between which is found the 8200 CalBP δ^{18}O precipitation event in the core. Notwithstanding the clear dating problem with the core in the area between depths 408cm and 450cm (see Fig. 11), much of the supposed two-hundred-year offset no longer affects the 8200 CalBP δ^{18}O precipitation event. If the two outlier ages for depths 408cm and 450cm are excluded, and the assumed dating resolution is reduced to approximately three dates per meter of core (based on twenty-one dates over 7.2m of core), the ages for the δ^{18}O precipitation event now end at 640cm, with a value of 8336–8210 CalBP at 68.2% probability (with the slightly more likely sub-range at 8264–8210 CalBP; 8359–8193 CalBP at 95.4% probability), and start at 647cm depth with a value of 8349–8231 Cal BP at 68.2% probability (8370–8205 CalBP at 95.4% probability: see Figure 12). In short, the calibrated ages for the terrestrial signal observed by Daley et al. (2009) (see Fig. 12) can be regarded as possibly just about contemporary with the 8200 CalBP event, with no requirement for a two-hundred-year offset, especially if the 8200 CalBP date itself is allowed to have a small dating error or range.

Similar dating issues arise with another major rapid climate episode in the Holocene, the 2200 BC (c. 4200 CalBP) event,[4] as the shape of the radiocarbon calibration curve here does not facilitate narrowly defined dating (Fig. 13). Only an intensively sampled stratigraphic sequence and Bayesian analysis

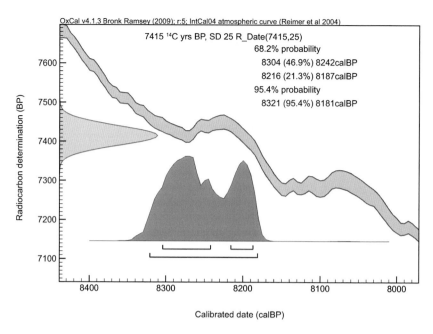

OxCal v4.1.3 Bronk Ramsey (2009); r:5; IntCal04 atmospheric curve (Reimer et al 2004)

Figure 8. The calibrated calendar age probability distribution in Calendar Years Before Present (CalBP) for a radiocarbon age with ±25 ^{14}C years measurement error exactly matching 8200 CalBP (thus: 7415±25 ^{14}C years BP). The assumption would be a short-lived sample (i.e., sub-annual to annual growth period, such as a seed) from a secure context (and so precisely) dating the year 8200 CalBP. Data from OxCal (Bronk Ramsey 1995, 2009) and IntCal04 (Reimer et al. 2004).

could hope to offer a fairly tight age for the event (Fig. 14). Moreover, both the 8200 CalBP and the 4200 cal yr BP events, because they are known from other high-resolution palaeoclimate sources, risk "absorbing" neighboring events and histories that have different real calendar age ranges; for example, events or contexts dating 8300–8100 CalBP, or 2300–2100 BC, can appear potentially to be compatible with 8200 CalBP or 2200 BC—but need not in reality be so. This can lead to a suck-in and smear effect (Baillie 1991), aggregating what are really distinct events and observations.

The outcome of this review of the properties of radiocarbon calibration is that the overall "quantum" ^{14}C chronology (as in Fig. 6) indicates that dating is preferentially attracted to the long steep slopes in the radiocarbon calibration curve. And, as illustrated by discussions of two well-known cases of rapid climate change considered important to the archaeological-palaeoenvironmental record, periods with wiggles and inversions involve the reverse situation, and pose challenges in many cases for high-resolution dating.

There are routes to overcoming the biases or limitations of the calibration curve. In particular, use of Bayesian analytical approaches that combine prior knowledge (e.g., a series order or other relationships) with the radiocarbon

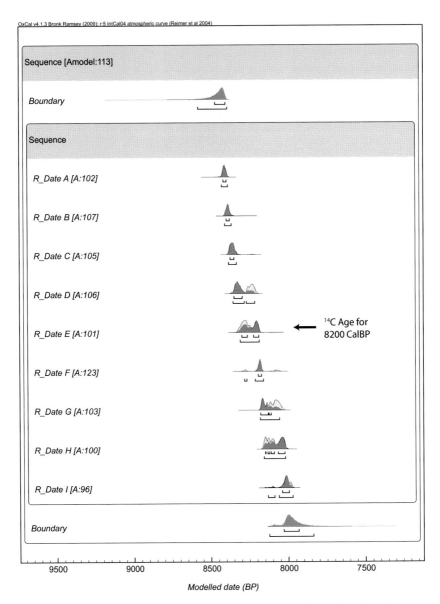

OxCal v4.1.3 Bronk Ramsey (2009); r:5 IntCal04 atmospheric curve (Reimer et al 2004)

Sequence [Amodel:113]

Boundary

Sequence

R_Date A [A:102]

R_Date B [A:107]

R_Date C [A:105]

R_Date D [A:106]

R_Date E [A:101] ← ¹⁴C Age for 8200 CalBP

R_Date F [A:123]

R_Date G [A:103]

R_Date H [A:100]

R_Date I [A:96]

Boundary

9500   9000   8500   8000   7500

Modelled date (BP)

Figure 9. Bayesian analysis using the **Sequence** analysis function in OxCal (Bronk Ramsey 1995, 2009) with IntCal04 (Reimer et al. 2004) of the hypothetical units A to I stratigraphic sequence (see text). Dating resolution assumed to be ±18 ¹⁴C years. Unit E—given the radiocarbon age for 8200 CalBP from IntCal (thus a "perfect" dating result in this hypothetical exercise)—is calculated as 8296-8260 CalBP (31.3%) and 8220–8187 CalBP (36.9%) at 68.2% probability and 8306–8184 at 95.4% probability. Thus the modelled calendar ages include the correct age (8200 CalBP), but the shape of the calibration curve across this period nonetheless spreads out the calendar age calculated into a 122-year range at 95.4% probability. For discussion and review of Bayesian analysis applied to radiocarbon, see Buck (2004); Bronk Ramsey (1995, 2009); Bayliss (2009); Blaauw et al. (2007).

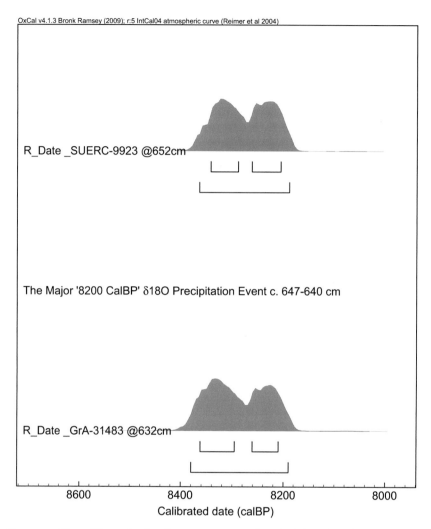

OxCal v4.1.3 Bronk Ramsey (2009); r:5 IntCal04 atmospheric curve (Reimer et al 2004)

R_Date _SUERC-9923 @652cm

The Major '8200 CalBP' δ18O Precipitation Event c. 647-640 cm

R_Date _GrA-31483 @632cm

| 8600 | 8400 | 8200 | 8000 |

Calibrated date (calBP)

Figure 10. The calibrated calendar age ranges and probability distributions for the two radiocarbon dates from Core NDN02/1 before and after the δ¹⁸O precipitation event at depth 640–647cm linked by Daley et al. (2009) with the "8200 Cal BP cold event": GrA-31483 at 632cm depth, 7475±50BP and SUERC-9923 at 652cm depth, 7460±40BP (data from ftp://rock.geosociety.org/pub/reposit/2009/2009203.pdf [last accessed 05 April 2010]). The very ambiguity noted in the text above and illustrated in Figure 8 is clearly present in both dates. Either or both could date at, or very close to, 8200 CalBP. But, because of the range of possible calendar ages available from these dates, and among the other 21 ¹⁴C dates from the core, the modelled age-depth curve (using the mid-points of the calibrated ranges) employed by Daley et al. (2009) ends up placing the 647–640cm depth period at 8338–8391 CalBP.

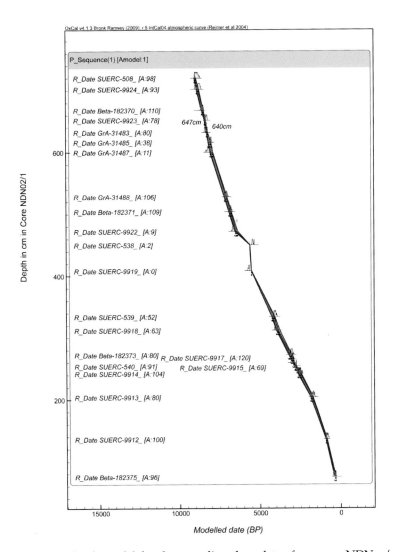

Figure 11. Age-depth model for the 23 radiocarbon dates from core NDN02/1 in Daley et al. (2009; from ftp://rock.geosociety.org/pub/reposit/2009/2009203.pdf [last accessed 05 April 2010]) using OxCal (Bronk Ramsey 2008) with an assumed 1cm sampling resolution and IntCal04 (Reimer et al. 2004). The non-modelled calibration histogram for each date is light grey and the modeled age is in dark grey. The A value in brackets for each date indicates the OxCal agreement index. Values below 60 indicate a lack of agreement between the modelled age and the non-modelled calibrated age at approximately the 95% level. Two data (SUERC-538 and SUERC-9919) especially are entirely inconsistent, and there would appear to be a dating issue for the core around 450–408cm depth. Thus the overall model has a poor agreement score. The calculated (modelled) date ranges for depths 640cm and 647cm (the end and start of the 8200 CalBP δ¹⁸O precipitation event) are 647cm 8338–8291 CalBP at 68.2% probability and 8357–8268 CalBP at 95.4% probability, and 640cm 8290–8240 CalBP at 68.2% probability and 8313–8218 Cal BP at 95.4% probability.

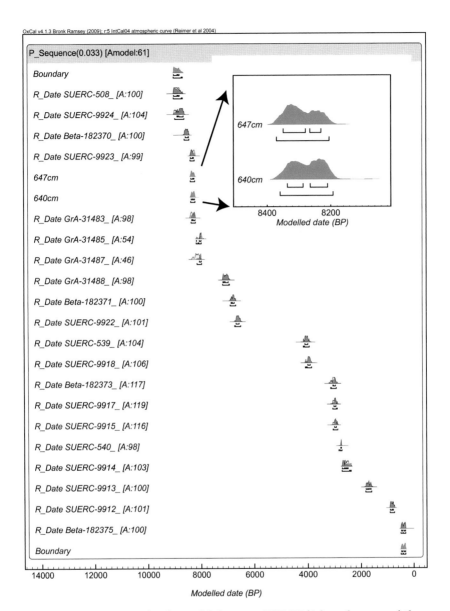

Figure 12. Revised age-depth model for core NDN02/1 based on 21 of the 23 radiocarbon dates (excluding SUERC-538 and SUERC-9919) in Daley et al. (2009) employing OxCal (Bronk Ramsey 2008) with an assumed c. 33cm sampling resolution and IntCal04 (Reimer et al. 2004). The light-gray histograms show the non-modelled calibration for each date, and the modeled age is shown in dark grey (the lines under each histogram indicate the 68.2% and 95.4% probability; dating ranges, respectively). The A value in brackets for each date indicates the OxCal

continued

Figure 12, *continued*

agreement index. Values below 60 indicate a lack of agreement of modelled age with the non-modelled calibrated age at approximately the 95% level. Despite two slight outliers, the overall model shown has a satisfactory agreement level (A = 61). The calculated date ranges for depths 640cm and 647cm (the end and start of the 8200 CalBP δ¹⁸O precipitation event) are shown in more detail in the inset. The age ranges are 647cm (start event) 8349–8279 CalBP (49.2%) and 8265–8231 CalBP (19%) at 68.2% probability and 8370–8205 CalBP at 95.4% probability, and 640cm (end event) 8336–8286 (31.4%) and 8264–8210 Cal BP (36.8%) at 68.2% probability and 8359–8193 CalBP at 95.4% probability.

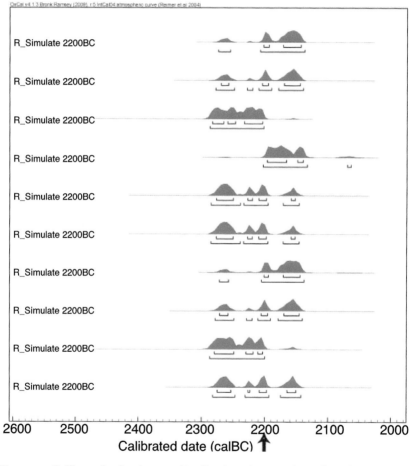

Figure 13. Calibrated calendar age distributions for 10 radiocarbon dates simulating the calendar date of 2200 BC with a ¹⁴C dating error of just ±10 years (thus I am assuming each date is the weighted average of several good-quality AMS radiocarbon dates). The upper and lower lines under each histogram show the 1σ and 2σ ranges, respectively. The specific date of 2200 BC is not clearly resolved and dating spreads unevenly over a two century time range. Data from OxCal (Bronk Ramsey 1995, 2009) and IntCal04 (Reimer et al. 2004).

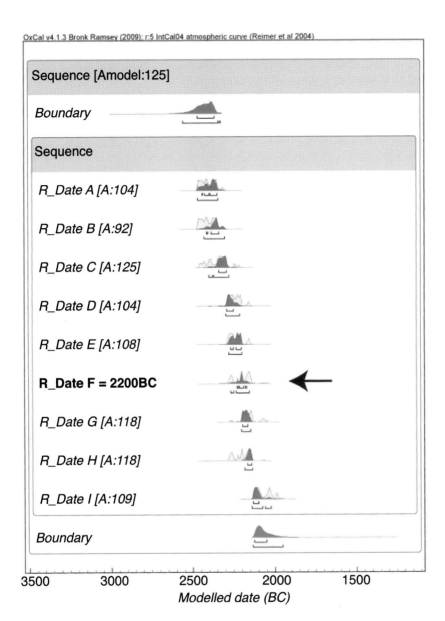

Sequence [Amodel:125]

Boundary

Sequence

R_Date A [A:104]

R_Date B [A:92]

R_Date C [A:125]

R_Date D [A:104]

R_Date E [A:108]

R_Date F = 2200BC

R_Date G [A:118]

R_Date H [A:118]

R_Date I [A:109]

Boundary

3500   3000   2500   2000   1500

Modelled date (BC)

Figure 14. A hypothetical archaeological (or stratigraphic sequence) "best case" dating for the 2200 BC event. Nine stratigraphic units are assumed, and best quality directly relevant short-lived samples (Boaretto 2009) are radiocarbon dated with pairs yielding pooled dates for each unit at ±18 ^{14}C years BP. Unit F dates to 2200 BC and, because (e.g., micromorphological indications) units above and below this unit were intensively sampled, five units cover about the 100–calendar-year period surrounding 2200 BC. To further make this a "perfect" case, the dates for each unit are taken from the tree-ring record of IntCal98 (Stuiver et al. 1998b) except 2200 BC, which has the value (±18 ^{14}C years BP) from IntCal04 itself (Reimer et al. 2004); so Unit A (2405 BC) 3905±18 ^{14}C years BP, Unit

continued

Figure 14, *continued*

B (2355 BC) 3905±18 ¹⁴C years BP, Unit C (2305 BC) 3859±18 ¹⁴C years BP, Unit D (2255 BC) 3804±18 ¹⁴C years BP, Unit E (2225 BC) 3799±18 ¹⁴C years BP, Unit F (2200 BC) 3777±18 ¹⁴C years BP, Unit G (2175 BC) 3751±18 ¹⁴C years BP, Unit H (2155 BC) 3777±18 ¹⁴C years BP, and Unit I (2105 BC) 3671±18 ¹⁴C years BP. Such a "best case" analysis does much more closely (and correctly) resolve the 2200 BC date, but still within ranges 2228–2221 BC (5.5%), 2213–2188 (56.2%), and 2178–2170 BC (6.3%) at 68.2% probability and 2268–2249 BC (6.2%) and 2234–2154 BC (89.2%) at 95.4% probability. The most likely sub-range within the overall 95.4% probability range is an 80-year calendar range nonetheless. Thus we obtain good but not extremely high-resolution dating, and it would be a challenge to obtain dating information as good as this hypothetical example from most archaeological or palaeoenvironmental contexts. Data and analysis are from OxCal (Bronk Ramsey 1995, 2009) and IntCal04 (Reimer et al. 2004). The hollow histograms show the calibrated ranges for each radiocarbon age in isolation and the solid histograms show the modelled range applying the Bayesian analysis. The A value in brackets for each date indicates the OxCal agreement index. Values above 60 indicate agreement of modelled age with the non-modelled calibrated age at approximately the 95% level. The lines under each histogram indicate the modelled 68.2% probability and 95.4% probability ranges, respectively.

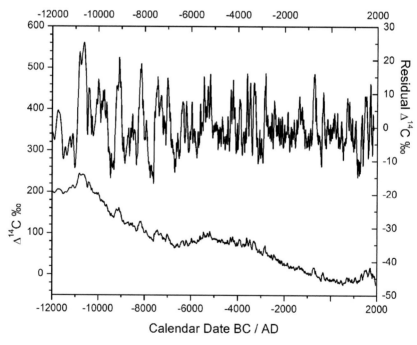

Figure 15. Bottom: The Δ¹⁴C record from IntCal04 in ‰ (Δ is age-corrected following Stuiver and Polach 1977), data from Reimer et al. (2004). Top: The residual Δ ¹⁴C record after removing the 1000-year moving average (so IntCal04 Δ ¹⁴C ‰ minus 1000-year moving average), data from Reimer et al. (2004).

data as employed in Figures 9, 11, 12, and 14. In a perfect world a tree-ring sample with bark might be tied to a specific context or event, and the ^{14}C dating of a dendro-defined sequence of these tree rings and the "wiggle-matching" of these to the calibration curve (Bronk Ramsey et al. 2001; Galimberti et al. 2004; Kromer 2009) might yield even greater dating precision. Such wiggle-match dating can also be included in a larger Bayesian analysis of a stratigraphic sequence. In such cases resolution more at the decadal or few decades scale might be hoped for, witness recent efforts to date the major volcanic eruption of the Santorini (Thera) volcano in the second millennium BC by [14]C (Manning et al. 2006; Friedrich et al. 2006), although, even then, success may be only partial unless circumstances are favorable (see Figs. 9 and 14; also see Blaauw et al. 2007).

In light of their prominence in the practice of radiocarbon dating, what do the major steep slopes in the ^{14}C calibration curve mean? The slopes indicate increased ^{14}C production in the atmosphere. We can see these changes in atmospheric ^{14}C best by looking at the Δ^{14}C record, especially once the longer-term trend is removed (Fig. 15).

On the basis of simplified models of atmospheric and atmosphere-ocean exchange processes (which lead to damping of the signal and a short lag between ^{14}C production and atmospheric observation), various estimates for atmospheric ^{14}C production based on the changing ^{14}C levels have been made. One such model of ^{14}C production (the iterative model described by Usoskin and Kromer 2005) is shown against the residual Δ^{14}C record in Figure 16.

RADIOCARBON, SOLAR ACTIVITY, AND CLIMATE

What causes these production peaks? As mentioned earlier, for the Holocene period, the major forcing is the variation in solar activity across this timescale. We can see this when we consider recent centuries where there are observations of solar activity variations in the form of a record of sunspot numbers (Hoyt and Schatten 1998a, 1998b). The variations in the radiocarbon record (the Δ^{14}C record) correlate closely with the observed sunspot activity especially allowing for a short lag between ^{14}C production and the tree-ring record (Fig. 17).

Little or no sunspot activity correlates with increased ^{14}C production, and higher sunspot activity correlates with reduced ^{14}C production. This correlation has wider relevance to climate. The Maunder Minimum period (AD 1645–1715) saw little or no sunspot activity and is closely correlated with some of the colder temperatures of the last 350 years on Earth (Bradley and Jones 1993; Mann et al. 2008; Jones et al. 2009, fig. 7). This association, and the consistent correlations apparent for the other recent atmospheric ^{14}C production peaks, led scholars to propose a link between times of low solar activity and climate anomalies, and the reverse (Eddy 1977; Hoyt and Schatten 1997; Crowley and Kim 1996). And, in general, despite debate, incomplete knowledge, and some skepticism, work in the last couple of decades based on a variety of measures and approaches indicates some level of significant forcing of climate change

on Earth because of changing solar activity.[5] Recent debate has focused on the issue (or problem) of the mechanism(s): exactly how the observed variations in solar activity force climate change on Earth.[6] Overall, the noticeable observation is the major shift in general opinion in the last decade or so; attempts into the early 1990s even to link changing levels of ^{14}C (and hence the sun) with evidence of past climate anomalies (e.g., Magny 1993) were not accepted unanimously, in part due to insufficient age control of the climate data, and in part because, with one exception (Stuiver et al. 1998a), the major ^{14}C records were constructed on a ten- to twenty-year average basis and thus for circular reasons may have tended to conform to a strong eleven-year (Schwabe sunspot) solar cycle.

On the longer time scale of the last several thousand years especially, the evidence has been shown to be much stronger linking sun and climate in work published over the last decade. For example, from high-resolution marine sediments in the North Atlantic documenting iceberg advance, there is evidence of nine widespread cooling episodes in the Holocene, starting at 10400 cal BP (8900 BC), and with the most recent one synchronous to the general cooling in mid- and high latitudes of the Northern Hemisphere between the fourteenth and eighteenth centuries AD. These century-scale episodes of iceberg advances into the North Atlantic are in excellent synchrony to major ^{14}C maxima (Bond et al. 2001) and other proxies indicating cool periods (e.g., glacier advances; Bradley et al. 2003). Work on lake sediments from southwest Alaska has yielded consistent evidence indicating that this is likely a hemisphere-wide pattern (Hu et al. 2003)—and, although subsequent work on ice-rafting provides increased resolution, and indicates some ambiguities, it nonetheless identifies many of the same key episodes (Moros et al. 2006). Bond et al. (2001) speculate that, even though the solar variations are relatively small, they can force alterations in the Atlantic Oscillation/North Atlantic Oscillation (Shindell et al. 2001), which, in turn, can have hemisphere-wide consequences. Detailed peat sequence analysis also closely links deterioration in climate during Holocene episodes with sharp changes in atmospheric ^{14}C and so solar activity.[7] Hence, the record of strong ^{14}C production changes (the big steep slopes in the radiocarbon calibration curve) does appear to mark periods of solar-induced climate change; the effects of such changes will of course vary depending on location on Earth and the underlying climate state.

Perhaps the main outstanding issue is that the calculated changes in solar activity appear too small to explain the climate shifts on Earth by themselves (see note 6); hence, amplifying mechanisms within the Earth's climate systems, whether general, or regionally applicable, are necessary if the sunclimate association is to be accepted and useful (Lean and Rind 2001). A variety of recent work has begun to provide models that can plausibly magnify and regionalize solar forcing consistent with observational or model data (e.g., Meehl et al. 2009; Renssen et al. 2006; Weber et al. 2004; Haigh 2003, 1996; Shindell et al. 2001; Marsh and Svensmark 2000).

Recent and still contentious work may indicate an additional relevant forcing in need of more careful attention: there is an intriguing correlation

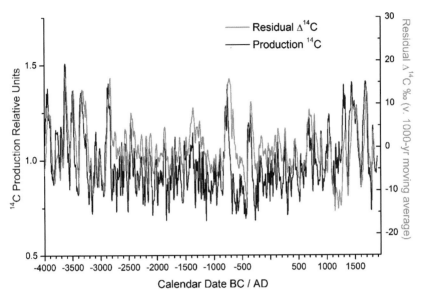

Figure 16. ¹⁴C production and change 4000 BC to AD 1900. Residual Δ¹⁴C record from the IntCal04 dataset (Reimer et al. 2004) after removing the 1000-year moving average (see Fig. 15). An approximation of ¹⁴C production is calculated using a standard box model of the carbon cycle (Siegenthaler et al. 1980) based on the assumption that oceanic ¹⁴C uptake was constant during this time, as described by Usoskin and Kromer (2005, 32) from the decadal tree-ring data of IntCal98 (Stuiver et al. 1998b); data courtesy of Bernd Kromer. The main features are the varying attenuation, and small phase shifts, between production and the signal recorded in the tree rings (a product of the carbon cycle and the frequency of production variations).

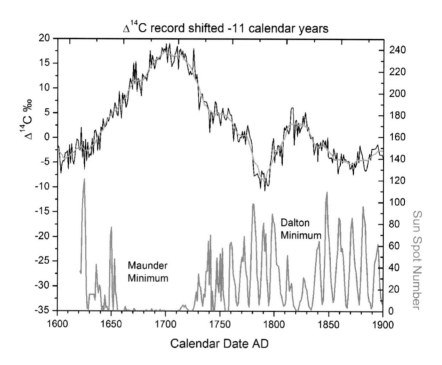

Figure 17. The annual Δ¹⁴C record AD 1600–1900 (Stuiver et al. 1998a) (black) and an 11-year moving average of this record (cyan), both shifted –11 calendar years (a rough allowance for the lag at this period between ¹⁴C production and the record found in trees) plotted against the historical records of sunspot numbers (SSN) (Hoyt and Schatten 1998a, 1998b) from http://www.ngdc.noaa.gov/stp/SOLAR/ftpsunspotnumber.html [last accessed 05 April 2010]. Few or no sunspots (Maunder Minimum, Dalton Minimum) correlate with a relatively quiet sun and generally cooler conditions on earth (Eddy 1976, 1977; Hoyt and Schatten 1997) and with increased ¹⁴C production, and vice versa.

between records of apparent century-scale archaeomagnetic jerks and several of the major climate change episodes of the last several millennia,[8] and especially the significant cooling episodes correlated with reduced solar activity and increased cosmic-ray–induced production of ^{14}C and ^{10}Be as identified by Bond et al. (2001). The suggestion is that such geomagnetic shifts, for example, via a tilting of the dipole to lower latitudes, may have enhanced cosmic-ray–induced nucleation of clouds (Courtillot et al. 2007)—perhaps especially at lower latitudes (Knudsen and Riisager 2009). If correct (contrast Bard and Delaygue 2008 with Courtillot et al. 2008), this would potentially provide another important magnifying feedback mechanism to increase the impact of reduced or increased solar activity on Earth.

RADIOCARBON, CLIMATE, AND PATTERN IN THE ARCHAEOLOGICAL/HISTORICAL RECORD?

If we integrate the information reviewed above, we find two key points: (1) the major steep slopes in the radiocarbon calibration curve—^{14}C production peaks (Figs. 15 and 16)—are much more likely to yield dates in the archaeological record as a whole (Figs. 6 and 7); and (2) these same major steep slopes in the radiocarbon calibration curve correlate with a series of major (cooling) climate-change episodes through the Holocene (cf. Bond et al. 2001). We might further suppose times of greatly reduced ^{14}C production (major troughs) likely correlate with warming periods, and, since these are inherently associated with inversions (or wiggles) in the radiocarbon calibration curve, the radiocarbon dating of these (or other wiggle/inversion periods on the radiocarbon calibration curve) will be more problematic as there will often be multiple and spread out dating probabilities (e.g., Figs. 4, 5, 8, 9 and 13).

The impact of the major cooling episodes would be very different in different areas of the Earth (I am restricting this chapter to the Northern Hemisphere). For example, such solar-driven cooling is thought to lead to a cyclonic circulation and higher frequencies of cold polar air in the North Atlantic and northwestern Europe (Hurrell 1995) and thus to a less favorable (cold, wet) climate in Northern to Central Europe. But, at the same time, the eastern Mediterranean to Near East region will be in anti-phase, and (once into the established post-glacial Holocene climate state) vegetation in the Mediterranean may have enjoyed favorable moisture and temperature conditions leading to early and longer growth seasons (Lean and Rind 1998); Vita-Finzi (2008) has recently suggested that the Mediterranean region received an (agriculturally beneficial) increase in small, non-erosive rains during such periods of solar cooling (and van Geel et al. [2004] observe positive effects of the ninth–eighth-century BC episode with regard to the Scythian culture). Thus, once properly out of glacial conditions and into the Holocene climate stage, it might be predicted (in general) that the major steep slopes in the radiocarbon calibration curve would correspond to periods of development or positive change in the archaeological-historical record in the eastern

Mediterranean–Near East, and that these episodes will stand out in overall dating chronologies (and, in contrast, at the end of the glacial period, and during the initial Holocene, one might predict the importance of warmer conditions, and at that time therefore wetter conditions, in the east Mediterranean region which likely correspond with plateau periods on the radiocarbon calibration curve [Manning et al. 2010]). Hence, they will likely be over-represented in our retrospective archaeological-historical assessments where radiocarbon is a key dating framework. In areas where cooling is linked to negative impacts (cool and humid phases), like Northern and Central Europe, we might expect the reverse: well-dated downturns or less positive changes. Apart from the Little Ice Age, perhaps the best known of these episodes is the circa ninety-year major solar minimum event centered around 765 BC associated with cool and humid phases in Central and Northern Europe (van Geel et al. 1998) and a horizon of climate/environmental and cultural change around the world (van Geel et al. 1996, 2004; Renssen et al. 2006; Chambers et al. 2007).

Do these generalizing observations manifest themselves if we look at the past of the east Mediterranean and Near East region? Of course any attempt to correlate time, climate, and human history is fraught with problems and complexities; climate does not create history—rather at most it creates contexts that may promote or not promote certain courses. The most recent attempt at a comprehensive review of global mid- and late Holocene climate can be found in Wanner et al. (2008; for critical review of the last one thousand years, see also Jones et al. 2009)—and one of the issues Wanner et al. highlight is the difficulty (i.e., absence) of any clear rapid or dramatic climate transition in even a majority of the climate proxy time series they review (p. 1818). Nonetheless, if we consider the ^{14}C-based solar indicators (above) and those proxies relevant to the Mediterranean region (Rosen 2007, especially the speleothem evidence from Soreq Cave; an important new speleothem record from Sofular Cave in northern Turkey [Fleitmann et al. 2009], adds new information), we can, despite the inevitable complications and complexity if any case is examined in detail, find some resonance at the gross and approximate (and cursory) scale (e.g., Fagan 2004; see also Fig. 18).

Figure 18 shows the radiocarbon calibration curve and residual Δ ^{14}C record for the last five thousand years (for one model of production, see Fig. 16). I have marked several periods for brief and sketched attention in light of the discussion above:

A. The century-scale major ^{14}C production peak centered c. 2860 BC approximately marks the beginning of major or important and primarily radiocarbon dated Early Bronze Age cultural phases in the Aegean region (Manning 1995; Kouka 2009), and, broadly, c. 3000/2900 BC, or the thirtieth century BC, is notable as a date for new cultural or socio-political-economic phases in the east Mediterranean and Near East (e.g., Dynasty 1 in Egypt, Wenke 2009; the Early Dynastic I period in Mesopotamia, Kuhrt 1995), and follows the overall cooling and very marked variability of the mid- to later fourth millennium BC and the production low and the likely more arid episode c. 3200–3000 BC—see Figure 16.[9] A long relatively stable

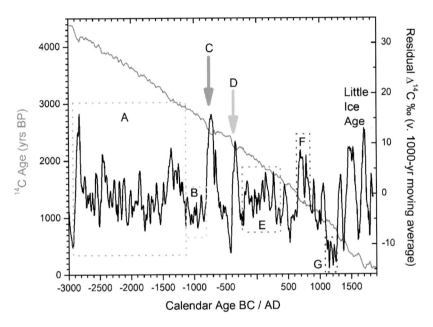

Figure 18. The radiocarbon calibration curve (IntCal04 from Reimer et al. 2004) in magenta versus the residual Δ ¹⁴C record with a 1000-year moving average removed (as Figs. 15 and 16) from Reimer et al. (2004) with several features identified: see text for discussion.

period until toward the end of the second millennium BC follows—this is the span more or less of the Bronze Age (Early, Middle, and Late) in this wider region (Kuhrt 1995; Mathers and Stoddart 1994).

B. There is a production low in the centuries either side of 1000 BC and this equates with the end/collapse of the Bronze Age (what used to be called a Dark Age) and major cultural decline and reorientation in the east Mediterranean and Near East.

C. The large century-scale ^{14}C production peak centered on 765 BC (cf. Bond et al. 2001, episode c. 800 BC) marks the take-off/florescence of the Iron Age and major positive developments in several regions of the east Mediterranean and Near East (and the reverse in Northern Europe)—see further in next section.

D. The ^{14}C record would indicate that attention should be paid to the contrasting situations of a major ^{14}C production trough centered c. 445 BC and then the substantial ^{14}C production peak centered c. 360 BC, and so some likely contrasting extremes all within a century. Climate-related information from ancient Greek sources is unfortunately less than consistent (Sallares 1991, 392–95), but could be relevant.

E. A notably stable period follows, and this more or less corresponds with the Hellenistic-Roman era (until some change/reorganization in the pattern begins in the 5th–6th centuries AD).

F. Much increased ^{14}C production c. AD 600–850 (and a major century-scale solar minimum c. AD 685) and a Bond et al. (2001) episode around AD 600 and several significant volcanic eruptions correspond to the so-called Migration Period and often less than ideal (cold, severe) winter circumstances in northern and western Europe (McCormick et al. 2007).

G. The major three-centuries-scale ^{14}C production trough here corresponds to the Medieval Climate Optimum (or Medieval Warm Period) best observed in Europe and the North Atlantic and for which varying dates are given according to different forms of analysis (Mann 2002; McCormick et al. 2007, 871 n. 20). The subsequent series of strong ^{14}C production peaks in the fourteenth to eighteenth centuries AD correspond to the Little Ice Age (Grove 1988; Bradley and Jones 1995; Fagan 2000).

THE EIGHTH CENTURY BC IN THE AEGEAN REGION

A major climate change episode starting c. AD 800 has been noted and widespread (global) impacts have been proposed.[10] This episode has been linked with a major solar minimum (see above). The Aegean region in the east Mediterranean offers an interesting case. There is a notable horizon of closely synchronous major (positive) cultural change across the wider Aegean region (and the entire Mediterranean region) in the eighth century BC, including the well-known Greek "Revolution" or "Renaissance" with its evidence for marked population growth, new widespread cultural and trade links (and Greek colonies from the eighth century BC), new wealth, the re-emergence of figural art, writing, and so on.[11] In central Turkey at Gordion the mid-eighth

century BC sees the construction of the great Midas Mound tumulus (dated c. 740 BC; Manning et al. 2001), and the other major tumuli at the site (W, P) likely also lie in the earlier to mid-eighth century BC. Muscarella (1995) proposed linking this (sudden) new wave of extraordinary monumentality with the creation of state-level society and kingship at Gordion (capital of the Phrygians)—following or during a period of dramatic social, economic, and political changes and/or pressures (both patterns in keeping with many other early state societies; Trigger 1990, 2003). We might suggest that the century-scale major solar minimum—associated it seems also with a possibly magnifying archaeomagnetic jump (Gallett et al. 2005)—centered c. 800 BC thus provided a key context for positive economic and social and hence political change at this same time in the Aegean-Mediterranean region.

The scenario for the Aegean region would thus comprise both a longer and more reliable growing season—when compared to recent less than favorable twentieth century AD ethnographic observations or the less than favorable later Classical Greek to Roman indications (e.g., as summarized in Garnsey 1988). This eighth century BC climate change may therefore have been, and kicked off, a boon period for the Aegean-Mediterranean region, and may have created conditions that actively promoted development and change in human societies, thanks to cooler climate, with more and more reliable rainfall, and so reduced interannual variability in harvests. Such circumstances could promote marked population growth in particular, and this in turn could create significant new pressures on social, political, ideational, technological, and subsistence systems within and between (adjacent) social groupings that could provide a context for, or promote, the rapid emergence of new leaders and political formations as suggested by Muscarella (1995) in the case of Gordion.

CONCLUSIONS

The relationships between radiocarbon dating and episodes of significant climate change during the Holocene have been revealed as complex and interwoven. Radiocarbon dating is not neutral. Major solar-forced cooling episodes are likely to be clearly (perhaps too clearly) represented, whereas some other types of events are more problematic in terms of tight dating resolution by radiocarbon. In both cases, for opposite reasons, there can be a tendency for a suck-in and smear effect to apply in scholarship: (1) around steep slopes in the radiocarbon calibration curve because of concentrated calendar probability making these times very pronounced and apparently important; but also, in reverse, (2) on wiggles and plateaus in the radiocarbon calibration curve where the wide calendar age ranges can seem to allow associations either with other similarly coarsely dated contexts, or with other events precisely dated by other means, when no real association may exist. Care and critique are thus necessary.

There is even a further small complication for the major solar minimum = ^{14}C production peak = major steep slopes in the radiocarbon calibration curve periods: at these times there are indications that some small regional (growing

season related) ^{14}C offsets exist when comparing the general Northern Hemisphere radiocarbon record (from trees in northern-central Europe and North America) against organic samples growing farther south in the east Mediterranean (Kromer et al. 2001; Reimer 2001; Kromer et al. 2010), making the high resolution of real calendar ages slightly problematic in, for example, the east Mediterranean at such times without use of a regional ^{14}C calibration curve (or adjustment).

At the same time, the radiocarbon record itself, derived from the underlying tree-ring time series, is an extraordinary and highly time-resolved solar proxy and thus solar-climate forcing proxy. This record offers a broad framework within which to consider profitably not only dating (and its problems), but also the long-term relationships among sun, climate, and history.

NOTES

1. For introductions to radiocarbon (^{14}C) dating, and especially as it is relevant to archaeology, Quaternary science, and even recent palaeoenvironmental sequences, see, e.g., Taylor (1997); Walker (2005: 17–53); and Turetsky et al. (2004: 326–32). An excellent Web site introduction to ^{14}C by Thomas Higham can be found at http://www.c14dating.com/ [last accessed 05 April 2010].

2. The brief summary in the main text is presented here more fully and supported with some references to the scientific literature for readers interested in some details:

Radiocarbon is a radioisotope produced in the upper atmosphere when galactic cosmic rays enter the Earth's atmosphere and react with ^{14}N atoms (Masarik and Beer 1999)—excepting very modern anthropogenic sources such as atmospheric nuclear explosions. ^{14}C is then oxidized to ^{14}CO$_2$ and mixes in the lower atmosphere forming a very small part (^{14}C comprises only about 1 out of every 10^{12} carbon atoms) of the overall atmospheric CO$_2$ reservoir. It was originally assumed that production was a constant, and that the ^{14}C activity of the atmosphere was in equilibrium with the biosphere and oceans. However, since the work of de Vries (1958, 1959), we know that production varies over time (for the Holocene—known clearly since Suess 1965—and beyond). These variations are caused by several factors (Damon and Sonett 1991; Stuiver et al. 1991). Modulation of the galactic cosmic ray flux reaching the Earth due to changes in solar magnetic activity is usually held to form the main shorter-term variable during the last 10,000 years: i.e., during periods of high solar activity distortion of the Earth's magnetic field by solar winds prevents (deflects) more galactic cosmic rays from entering the atmosphere, and the reverse (Stuiver and Quay 1980; Stuiver and Braziunas 1988, 1989, 1998; Stuiver et al. 1991). However, over longer-term timescales the Earth's geomagnetic field is seen as the key controlling regime (Sternberg 1992), with major changes therein particularly relevant—for example, the Mono Lake and Laschamp events around 34,000 and 40,000 years ago have been much discussed (e.g., Lal and Charles 2007) (on geomagnetic excursions, see Roberts 2008).

But, on shorter timescales and for the Holocene especially, the radiocarbon record is regarded as a key proxy record for variations in solar activity (Damon 1977; Stuiver and Braziunas 1989, 1993; Usoskin et al. 2005; Bard and Frank 2006)—confirmed by the good comparisons with records of another independent solar-modulated cosmogenic isotope, beryllium-10, ^{10}Be (Beer et al. 1988; Bard et al. 1997; Beer 2000; Beer et al. 2006). The main difference is that ^{10}Be is deposited at the Earth's surface much more quickly and directly after production in the atmosphere (typically 1–2 years) compared to ^{14}C, and is not damped by the carbon cycle or complicated by changes therein (in contrast with ^{14}C). Radiocarbon allows us to model solar activity variations in the past before observational or instrumental records exist (that is, before the seventeenth century AD). From observations of the recent few centuries, shorter-term cycles or variations in solar activity are known (11-year Schwabe cycle, 22-year Hale cycle, 88-year Gleissberg cycle [Hathaway et al. 1999; de Meyer 1998; Peristykh and Damon 2003]) and from the ^{14}C and ^{10}Be records longer-term cycles are observed: of around 205–210 years, and then, on longer multi-century timescales a record of major solar minima and maxima, with, perhaps most importantly, an apparent recurrent cycle of around 1500 years, which seems to be associated with key Holocene climate change episodes (Bond et al. 2001; Usoskin et al. 2004; van Geel et al. 1998; Hu et al. 2003; Wang et al. 2005).

However, the sun is not the only key factor in observed ^{14}C. The other major issue concerns the oceans (in the modern era there is also the impact of CO_2 from fossil fuels, the "Suess effect" [Suess 1955; Tans et al. 1979]). On the Earth, since the vast majority of ^{14}C (i.e., $^{14}CO_2$) ends up in the oceans, given the nature of the carbon cycle (Wigley and Schimel 2000), changes in exchange of CO_2 between the oceans and the atmosphere will also affect ^{14}C levels in the atmosphere, and here, too, research subsequent to the original conception of the ^{14}C method has shown that important variations occur (Stuiver et al. 1991; Broecker 1997). At the dramatic scale, the sharp rise in ^{14}C levels in the atmosphere associated with the onset of the Younger Dryas has been attributed to a (substantial) shut down in ocean overturning circulation (because of a huge release of meltwater into the north Atlantic; Broecker 2006 and refs.) at this time (Hughen et al. 2000; for discussion of the mechanisms involved in the radiocarbon changes, see Delaygue et al. 2003) (though cf. other scholars who suggest the primary forcing was solar; Renssen et al. 2000); at the more minor scale, patterns evident in the radiocarbon record following major cooling or warming episodes imply ocean modulation (as do the changing offsets—smaller during cooler periods, larger during warmer periods—between the Northern and Southern Hemisphere records; see Figs. 1–2).

For the history and development of radiocarbon calibration, see Taylor et al. 1996; Blackwell et al. 2005; Bronk Ramsey et al. 2006; Scott and Reimer 2009. For radiocarbon as a useful tracer of the way the global carbon cycle works, see Levin and Hesshaimer 2000.

3. There is a large body of literature on this event; see especially Rohling and Pälike 2005; Barber et al. 1999; Alley et al. 2005, 1997; Klitgaard-Kristensen et al. 1998; von Grafenstein et al. 1998; Gasse and Van Campo 1994. For the impact in the eastern Mediterranean, see Pross et al. 2009.

4. See, e.g,. Weiss et al. 1993; Weiss 2000, 78–91; Cullen et al. 2000; Marchant and Hooghiemstra 2004; Booth et al. 2006; Arz et al. 2006; Magny et al. 2009.

5. The literature on the association between solar activity changes and climate change on Earth is vast. Some scholars argue against a link; for example, Foukal et al. (2006). But in the last couple of decades there has been an increasing number of studies that have proposed or found some level of significant forcing of climate change on Earth as a result of changing solar activity. I give a selection of examples covering a range of approaches and data sources (and with further references): Knudsen and Riisager 2009; Wanner et al. 2008; Zhang et al. 2008; Bard and Delaygue 2008; Chambers et al. 2007; Renssen et al. 2000, 2006; Scafetta and West 2005; Le Mouël et al. 2005; Lean et al. 2005; 1995; Weber et al. 2004; Fleitmann et al. 2003; White et al. 2003; 1997; Solanki 2002; Neff et al. 2001; Bond et al. 2001; Shindell et al. 2001; Björk et al. 2001; Bard et al. 2000; Crowley 2000; Friis-Christensen 2000; Haigh 2000; 1996; Perry and Hsu 2000; deMenocal et al. 2000; van Geel et al. 1999; Lean and Rind 1998; Cliver et al. 1998; Hoyt and Schatten 1997; Crowley and Kim 1996; Damon and Jirikowic 1992; Mitchell et al. 1979. For the evidence available on the history of solar activity, an excellent resource is I. G. Usoskin, "A History of Solar Activity over Millennia" at http://www.livingreviews.org/lrsp-2008-3 [last accessed 05 April 2010]. Reconstructions of past solar irradiance beyond the instrumental record have been made on the basis of radionuclides. The most recent at the time of writing, and for the last 9300 years, employing ^{10}Be, is by Steinhilber et al. (2009). (The paper of Field et al. [2006] highlights, however, that it can be problematic to disentangle climate-driven changes versus solar-driven changes in the ^{10}Be record, and that solar signals may be overestimated. Hence, the comparison with ^{14}C-based reconstructions that— despite carbon-cycle damping and forcing—are not affected by weather [e.g., precipitation and wind at study locus as is the case for ^{10}Be] remains important.) The periods noted from the ^{14}C record in the text below stand out (Steinhilber et al. 2009, fig. 3) (mid-fourth millennium BC, early third millennium BC, around 800 BC, the major irradiance trough in the seventh century AD, and the Little Ice Age—especially Spörer Minimum).

6. Lean and Rind (2001); Rind (2002); Reid (2000); Friis-Christensen (2000). The key problem is that changes in solar activity appear several times too small as calculated irradiance changes to create the substantial climate shifts observed on Earth by themselves (Foukal et al. 2004, 2006)—see further in the main text.

7. For example, van Geel et al. (1998); Mauquoy et al. (2002, 2004); Blaauw et al. (2004); Chambers et al. 2007 (and with the last contrary to the questioning of the solar linkage in Plunkett [2006]).

8. Courtillot et al. (2007); Gallet et al. (2006, 2005); see also previously St-Onge et al. (2003)—see also the critique of Bard and Delaygue (2008) and the response by Courtillot et al. (2008), and the study of Knudsen and Riisager (2009) in general support of some linkage (and perhaps also of the linkage with, and among, cosmic rays, cloud formation, and climate). Others had previously proposed links between changes in global temperature on Earth and geomagnetic changes linked to solar activity (Cliver et al. 1998).

9. The later fourth millennium BC is a time of radical change in several areas of the world (Brooks 2006; Weiss 2000, 77). There is a general reorganization of settlement patterns in many areas and new cultural behaviors, structures, and larger socio-political entities develop especially in the last centuries—notably including a unified Egyptian state. The climate context (cooler and with increased aridity) seems important and was perhaps a driver in changes to increased social complexity in several areas (Brooks 2006). The African Humid Period ended fairly abruptly around 5500 years ago (3500 BC) (deMenocal et al. 2000) and ice-core evidence indicates a (presumably linked) sharp climate re-organization around 5200 years ago (3200 BC)—to the generally cooler late Holocene period (Thompson et al. 2006). There was a rapid decline in Northern Hemisphere solar insolation through the fourth millennium BC (e.g., Claussen et al. 1999, fig. 1). The radiocarbon record for the mid-fourth millennium BC is notable for a series of strong relatively closely spaced ^{14}C production peaks (centered about 3625 BC, 3500 BC, and 3335 BC) similar to the Little Ice Age period; thus we may suspect that an overall cooler solar activity context came to dominate, and speculate that the combination of these strong solar forcings seems to have passed a "tipping point" such that, via amplifying mechanisms or non-linear transfers, the changes in solar activity pushed the climate system into a rapid change c. 3500–3200 BC (deMenocal et al. 2000, 355–59). In such a general cooling context, the reduced ^{14}C production trough in the last couple of centuries of the fourth millennium BC (not unlike the Medieval Climate Optimum period) could then link with observations of an increased aridity episode for several centuries (in a generally cooler context) from about 3200 BC.

10. Van Geel et al. 1996, 1998, 2004; Renssen et al. 2006; Chambers et al. 2007.

11. See, for example, Snodgrass 1980; Hägg 1983; Morris 1987, 1998; Whitley 2001, esp. 98–101; Dickinson 2005, 5, 216–17; Hall 2007. The first use of the term "Renaissance" for the period from c. 770–700 BC in Greece was by Coldstream (1977, 2003). The suggestion of significant population growth centered in the eighth century BC was proposed by Snodgrass (1977, 1980, 1983), and finds support or feasibility in demographic analyses (Sallares 1991, 85–91). Ancient Greek inheritance norms may have promoted colonization given an increased birth rate (Hall 2007, 115–16).

References

Alley, R. B., P. A. Mayewski, T. Sowers, M. Stuiver, K. C. Taylor, and P. U. Clark. 1997. Holocene climatic instability: a prominent, widespread event 8200 yr ago. *Geology* 25: 483–86.

Alley, R. B., and A. M. Ágústsdóttir. 2005. The 8k event: cause and consequences of a major Holocene abrupt climate change. *Quaternary Science Reviews* 24: 1123–49.

Arz, H. W., F. Lamy, and J. Pätzold. 2006. A pronounced dry event recorded around 4.2 ka in brine sediments from the northern Red Sea. *Quaternary Research* 66: 432–41.

Baillie, M. G. L. 1991. Suck in and smear: two related chronological problems for the 90s. *Journal of Theoretical Archaeology* 2: 12–16.

Barber, D. C., A. Dyke, C. Hillaire-Marcel, A. E. Jennings, J. T. Andrews, M. W. Kerwin, G. Bilodeau, R. McNeely, J. Southon, M. D. Morehead, and J. M. Gagnon. 1999. Forcing of the cold event of 8,200 years ago by catastrophic drainage of Laurentide lakes. *Nature* 400: 344–48.

Bard, E., and G. Delaygue. 2008. Comment on "Are there connections between the Earth's magnetic field and climate?" by V. Courtillot, Y. Gallet, J.-L. Le Mouël, F. Fluteau, A. Genevey (*EPSL* [2007]: 253, 328). *Earth Planetary Science Letters* 265: 302–7.

Bard, E., and M. Frank. 2006. Climate change and solar variability: what's new under the sun? *Earth and Planetary Science Letters* 248: 1–14.

Bard, E., G. M. Raisbeck, F. Yiou, and J. Jouzel. 1997. Solar modulation of comogenic nuclide production over the last millennium: comparison between ^{14}C and ^{10}Be records. *Earth and Planetary Science Letters* 150: 453–62.

———. 2000. Solar irradiance during the last 1200 years, based on cosmogenic nuclides. *Tellus* 52B: 985–92.

Bayliss, A. 2009. Rolling out the revolution: using radiocarbon dating in archaeology. *Radiocarbon* 51: 123–47.

Beer, J. 2000. Long-term indirect indices of solar variability. *Space Science Reviews* 94: 53–66.

Beer, J., U. Siegenthaler, G. Bonani, R. C. Finkel, H. Oescheger, M. Suter, and W. Wölfli. 1988. Information on past solar activity and geomagnetism from ^{10}Be in the Camp Century ice core. *Nature* 331: 675–79.

Beer, J., M. Vonmoos, and R. Muscheler. 2006. Solar variability over the past several millennia. *Space Science Reviews* 125: 67–79.

Björck, S., R. Muscheler, B. Kromer, C. S. Andresen, J. Heinemeier, S. J. Johnsen, D. Conley, N. Koç, M. Spurk, and S. Veski. 2001. High-resolution analyses of an early Holocene climate event may imply decreased solar forcing as an important climate trigger. *Geology* 29: 1107–10.

Blaauw, M., and J. A. Christen. 2005. The problems of radiocarbon dating. *Science* 308: 1551.

Blaauw, M., B. van Geel, and J. van der Plicht. 2004. Solar forcing of climate change during the mid-Holocene: indications from raised bogs in the Netherlands. *The Holocene* 14: 35–44.

Blaauw, M., J. A. Christen, D. Mauquoy, J. van der Plicht, and K. D. Bennett. 2007. Testing the timing of radiocarbon events between proxy archives. *The Holocene* 17: 283–88.

Blackwell, P. G., C. E. Buck, and P. J. Reimer. 2005. Important features of the new radiocarbon calibration curves. *Quaternary Science Reviews* 25: 408–13.

Boaretto, E. 2009. Dating materials in good archaeological contexts: the next challenge for radiocarbon analysis. *Radiocarbon* 51: 275–81.

Bond, G., B. Kromer, J. Beer, R. Muscheler, M. Evans, W. Showers, S. Hoffman, R. Lotti-Bond, I. Hajdas, and G. Bonani. 2001. Persistent solar influence on North Atlantic surface circulation during the Holocene. *Science* 294: 2130–36.

Booth, R., S. T. Jackson, S. L. Forman, J. E. Kutzbach, E. E. Bettis, J. Krieg, and D. K. Wright. 2005. A severe centennial-scale drought in mid-continental North America 4200 years ago and apparent global linkages. *The Holocene* 15: 321–28.

Bradley, R. S., K. R. Briffa, J. Cole, M. K. Hughes, and T. J. Osborn. 2003. The climate of the last millennium. In *Paleoclimate, global change and the future*, ed. K. D. Alverson, R. S. Bradley, and T. F. Pedersen, 105–49. Berlin: Springer.

Bradley, R. S., and P. D. Jones. 1993. "Little Ice Age" summer temperature variations: Their nature and relevance to recent global warming trends. *The Holocene* 3: 367–76.

Broecker, W. S. 1997. Thermohaline circulation, the Achilles Heel of our climate system: will man-made CO_2 upset the current balance? *Science* 278: 1582–88.

Broecker, W. S. 2006. Abrupt climate change revisited. *Global and Planetary Change* 54: 211–15.

Bronk Ramsey, C. 1995. Radiocarbon calibration and analysis of stratigraphy: the OxCal program. *Radiocarbon* 37: 425–30.

———. 2008. Deposition models for chronological records. *Quaternary Science Reviews* 27: 42–60.

———. 2009. Bayesian analysis of radiocarbon dates. *Radiocarbon* 51: 337–60.

Bronk Ramsey, C., C. E. Buck, S. W. Manning, P. Reimer, and H. van der Plicht. 2006. Developments in radiocarbon calibration for archaeology. *Antiquity* 80: 783–98.

Bronk Ramsey, C., J. van der Plicht, and B. Weninger. 2001. "Wiggle matching" radiocarbon dates. *Radiocarbon* 43: 381–89.

Buck, C. E. 2004. Bayesian chronological data interpretation: where now? In *Tools for constructing chronologies*, ed. C. E. Buck and A. R. Millard, 1–24. London: Springer.

Chambers, F. M., D. Mauquoy, S. A. Brain, M. Blaauw, and J. R. G. Daniell. 2007. Globally synchronous climate change 2800 years ago: proxy data from peat in South America. *Earth and Planetary Science Letters* 253: 439–44.

Claussen, M., C. Kubatzki, V. Brovkin, and A. Ganopolski. 1999. Simulation of an abrupt change in Saharan vegetation in the mid-Holocene. *Geophysical Research Letters* 26: 2037–40.

Cliver, E. W., V. Boriakoff, and J. Feynman. 1998. Solar variability and climate change: geomagnetic aa index and global surface temperature. *Geophysical Research Letters* 25: 1035–38.

Coldstream, J. N. 1977. *Geometric Greece*. London: E. Benn.

———. 2003. *Geometric Greece*. Second edition. London: Routledge.

Courtillot, V., Y. Gallet, J.-L. Le Mouël, F. Fluteau, and A. Genevey. 2007. Are there connections between the earth's magnetic field and climate? *Earth and Planetary Science Letters* 253: 328–39.

———. 2008. Response to comment on "Are there connections between the earth's magnetic field and climate?, Earth Planet. Sci. Lett., 253, 328–339, 2007" by Bard, E., and Delaygue, M., Earth Planet. Sci. Lett., in press, 2007. *Earth and Planetary Science Letters* 265: 308–11.

Crowley, T. J. 2000. Causes of climate change over the past 1000 years. *Science* 289: 270–77.

Crowley, T. J., and K.-Y. Kim. 1996. Comparison of proxy records of climate change and solar forcing. *Geophysical Research Letters* 23: 359–62.

Cullen, H. M., P. deMenocal, G. Hemming, E. H. Brown, T. Guilderson, and F. Sirocko. 2000. Climate change and the collapse of the Akkadian Empire: evidence from the deep sea. *Geology* 28: 379–82.

Daley, T. J., F. A. Street-Perrott, N. J. Loader, K. E. Barberm, P. D. M. Hughes, E. H. Fisher, and J. D. Marshall. 2009. Terrestrial climate signal of the "8200 yr B.P. cold event" in the Labrador Sea region. *Geology* 37: 831–34.

Damon, P. E. 1977. Solar induced variations of energetic particles. In *The solar output and its variations*, ed. O. R. White, 429–48. Boulder: Colorado Associated University Press.

Damon, P. E., and C. P. Sonett. 1991. Solar and terrestrial components of the atmospheric ^{14}C variation spectrum. In *The sun in time*, ed. C. P. Sonett, M. S. Giampapa, and M. S. Matthews, 360–88. Tucson: University of Arizona Press.

Damon, P. E., and J. L. Jirikowic. Solar forcing of global climate change? In *Radiocarbon after four decades: an interdisciplinary perspective*, ed. R. E. Taylor, A. Long, and R. S. Kra, 117–29. New York: Springer.

de Jong, A. F. M., W. G. Mook, and B. Becker. 1979. Confirmation of the Suess wiggles: 3200–700 BC. *Nature* 280: 48–49.

Delaygue, G., T. F. Stocker, F. Joos, and G.-K. Plattner. 2003. Simulation of atmospheric radiocarbon during abrupt oceanic circulation changes: trying to reconcile models and reconstructions. *Quaternary Science Reviews* 22: 1647–58.

deMenocal, P., J. Ortiz, T. Guilderson, J. Adkins, M. Sarntheim, L. Baker, and M. Yarusinsky. 2000. Abrupt onset and termination of the African Humid Period: rapid climate responses to gradual insolation forcing. *Quaternary Science Reviews* 19: 347–61.

de Meyer, F. 1998. Modulation of the solar magnetic cycle. *Solar Physics* 181: 201–19.

de Vries, H. 1958. Variation in concentration of radiocarbon with time and location on earth. *Proceedings of the Koninklijke Nederlandse Akademie Van Wetenschappen Series B* 61: 94–102.

———. 1959. Measurement and use of natural radiocarbon. In *Researches in geochemistry*, ed. P. H. Abelson, 169–89. New York: John Wiley & Sons.

Dickinson, O. 2005. *The Aegean from Bronze Age to Iron Age.* London: Routledge.

Eddy, J. A. 1976. The Maunder Minimum. *Science* 192: 1189–1202.

———. 1977. Climate and the changing sun. *Climate Change* 1: 173–90.

Fagan, B. 2000. *The Little Ice Age: how climate made history 1300–1850.* New York: Basic Books.

———. 2004. *The long summer: how climate changed civilization.* New York: Basic Books.

Field, C. V., G. A. Schmidt, D. Koch, and C. Salyk. 2006. Modeling production and climate-related impacts on ^{10}Be concentration in ice-cores. *Journal of Geophysical Research* 111, D15107, doi: 10.1029/2005JD006410, 2006.

Fleitmann, D., S. J. Burns, M. Mudelsee, U. Neff, J. Kramers, A. Mangini, and A. Matter. 2003. Holocene forcing of the Indian monsoon recorded in a stalagmite from Southern Oman. *Science* 300: 1737–39.

Fleitmann, D., H. Cheng, S. Badertscher, R. L. Edwards, M. Mudelsee, O. M. Göktürk, A. Fankhauser, R. Pickering, C. C. Raible, A. Matter, J. Kramers, and O. Tüysüz. 2009. Timing and climatic impact of Greenland interstadials recorded in stalagmites from nothern Turkey. *Geophysical Research Letters* 36: doi:10.1029/2009GL040050.

Foukal, P., C. Fröhlich, H. Spruit, and T. M. L. Wigley. 2006. Variations in solar luminosity and their effect on the earth's climate. *Nature* 443: 161–66.

Foukal, P., G. North, and T. Wigley. 2004. A stellar view on solar variations and climate. *Science* 306: 68–69.

Friedrich, W. L., B. Kromer, M. Friedrich, J. Heinemeier, T. Pfeiffer, and S. Talamo. 2006. Santorini Eruption radiocarbon dated to 1627–1600 B.C. *Science* 312: 548.

Friis-Christensen, E. 2000. Solar variability and climate: a summary. *Space Science Reviews* 94: 411–21.

Galimberti, M., C. Bronk Ramsey, and S. W. Manning. 2004. Wiggle-match dating of tree ring sequences. *Radiocarbon* 46: 917–24.

Gallet, Y., A. Genevey, and F. Fluteau. 2005. Does earth's magnetic field secular variation control centennial climate change? *Earth Planetary Research Letters* 236: 339–47.

Gallet, Y., A. Genevey, M. Le Goff, F. Fluteau, and S. A. Eshraghi. 2006. Possible impact of the earth's magnetic field on the history of ancient civilizations. *Earth and Planetary Science Letters* 246: 17–26.

Garnsey, P. 1988. *Famine and food supply in the Graeco-Roman world: responses to risk and crisis.* Cambridge: Cambridge University Press.

Gasse, F., and E. Van Campo. 1994. Abrupt post-glacial climate events in West Asia and North Africa monsoon domains. *Earth and Planetary Science Letters* 126: 435–56.

Guilderson, T. P., P. J. Reimer, and T. A. Brown. 2005. The boon and bane of radiocarbon dating. *Science* 307: 362–64.

Hägg, R., ed. 1983. *The Greek renaissance of the eighth century BC: tradition and innovation.* Stockholm: Svenska Institut i Athen.

Haigh, J. D. 1996. The impact of solar variability on climate. *Science* 272: 981–84.

———. 2000. Solar variability and climate. *Weather* 55: 399–406.

———. 2002. The effects of solar variability on the earth's climate. *Philosophical Transactions of the Royal Society* A 361: 95–111.

Hall, J. M. 2007. *A history of the archaic Greek world ca. 1200–479 BCE.* Malden: Blackwell.

Hathaway, D. H., R. M. Wilson, and E. J. Reichmann. 1999. A synthesis of solar cycle prediction techniques. *Journal of Geophysical Research* 104: 22375–388.

Hoyt, D. V., and K. H. Schatten. 1997. *The role of the sun in climate change.* New York: Oxford University Press.

———. 1998a. Group sunspot numbers: a new solar activity indicator. Part 1. *Solar Physics* 181: 189–219.

————. 1998b. Group sunspot numbers: a new solar activity indicator. Part 2. *Solar Physics* 181: 491–512.

Hu, F. S., D. Kaufman, S. Yoneji, D. Nelson, A. Shemesh, Y. Huang, J. Tian, G. Bond, B. Clegg, and T. Brown. 2003. Cyclic variation and solar forcing of Holocene climate in the Alaskan subarctic. *Science* 301: 1890–93.

Hughen, K., J. R. Southon, S. J. Lehman, and J. T. Overpeck. 2000. Synchronous radiocarbon and climate shifts during the last deglaciation. *Science* 290: 1951–54.

Hurrell, J. W. 1995. Decadal trends in the North Atlantic Oscillation: regional temperatures and precipitation. *Science* 269: 676–79.

Jones, P. D., K. R. Briffa, T. J. Osborn, J. M. Lough, T. D. van Ommen, B. M. Vinther, J. Luterbacher, E. R. Wahl, F. W. Zwiers, M. E. Mann, G. A. Schmidt, C. M. Ammann, B. M. Buckley, K. M. Cobb, J. Esper, H. Goosse, N. Graham, E. Jansen, T. Kiefer, C. Kull, M. Küttel, E. Mosley-Thompson, J. T. Overpeck, N. Riedwyl, M. Schulz, A. W. Tudhope, R. Villalba, H. Wanner, E. Wolff, and E. Xoplaki. 2009. High-resolution palaeoclimatology of the last millennium: a review of current status and future prospects. *The Holocene* 19: 3–49.

Klitgaard-Kristensen, D., H. P. Sejrup, H. Haflidason, S. Johnsen, and M. Spurk. 1998. A regional 8200 cal. yr BP cooling event in northwest Europe, induced by final stages of the Laurentide ice-sheet deglaciation? *Journal of Quaternary Science* 13: 165–69.

Knudsen, M. F. and P. Riisager. 2009. Is there a link between Earth's magnetic field and low-latitude precipitation? *Geology* 37: 71–74.

Kouka, O. 2009. Third millennium BC Aegean chronology: old and new data from the perspective of the third millennium AD. In *Tree-rings, kings and old world archaeology and environment: papers presented in honor of Peter Ian Kuniholm*, ed. S. W. Manning and M. J. Bruce, 133–49. Oxford: Oxbow Books.

Kromer, B. 2009. Radiocarbon and dendrochronology. *Dendrochronologia* 27: 15–19.

Kromer, B., S. W. Manning, P. I. Kuniholm, M. W. Newton, M. Spurk, and I. Levin. 2001. Regional $^{14}CO_2$ offsets in the troposphere: magnitude, mechanisms, and consequences. *Science* 294: 2529–32.

Kromer, B., S. Manning, M. Friedrich, S. Talamo, and N. Trano. 2010. ^{14}C calibration in the 2nd and 1st millennia BC. Eastern Mediterranean Radiocarbon Comparison Project (EMRCP). *Radiocarbon*, in press.

Kuhrt, A. 1995. *The ancient Near East c.3000–330 BC.* London: Routledge.

Lal, D., and C. Charles. 2007. Deconvolution of the atmospheric radiocarbon record in the last 50,000 years. *Earth and Planetary Science Letters* 258: 550–60.

Lean, J., J. Beer, and R. Bradley. 1995. Reconstruction of solar irradiance since 1610: implications for climate change. *Geophysical Research Letters* 22: 3195–98.

Lean, J., and D. Rind. 1998. Climate forcing by changing solar radiation. *Journal of Climate* 11: 3069–94.

————. 2001. Earth's response to a variable sun. *Science* 292: 234–36.

Lean, J., G. Rottman, J. Harder, and G. Kopp. 2005. SORCE contributions to new understanding of global change and solar variability. *Solar Physics* 230: 27–53.

Le Mouël, J.-L., V. Kossobokov, and V. Courtillot. 2005. On long-term variations of simple geomagnetic indices and slow changes in magnetospheric currents. *Earth and Planetary Science Letters* 232: 273–86.

Levin, I., and V. Hesshaimer. 2000. Radiocarbon:—a unique tracer of global carbon cycle dynamics. *Radiocarbon* 42: 69–80.

Magny, M. 1993. Solar influences on Holocene climatic changes illustrated by correlations between past lake-level fluctuations and the atmospheric ^{14}C record. *Quaternary Research* 40: 1–9.

Magny, M., B. Vannière, G. Zanchetta, E. Fouache, G. Touchais, L. Petrika, C. Coussot, A.-V. Walter-Simonnet, and F. Arnaud. 2009. Possible complexity of the climate event around 4300–3800 cal. BP in the central and western Mediterranean. *The Holocene* 19: 823–33.

Mann, M. E. 2002. Medieval climate optimum. In *Encyclopedia of global environmental change. Volume 1: The earth system: physical and chemical dimensions of global environmental change*, ed. T. Munn, 514–16. Chichester: John Wiley & Sons.

Mann, M. E., Z. Zhang, M. K. Hughes, R. S. Bradley, S. K. Miller, S. Rutherford, and F. Ni. 2008. Proxy-based reconstructions of hemispheric and global surface temperature variations over the past two millennia. *Proceedings of the National Academy of Sciences of the United States of America* 105: 13252–57.

Manning, S. W., C. Bronk Ramsey, W. Kutschera, T. Higham, B. Kromer, P. Steier, and E. Wild. 2006. Chronology for the Aegean Late Bronze Age. *Science* 312: 565–69.

Manning, S. W., C. McCartney, B. Kromer, and S. T. Stewart. 2010. The earlier Neolithic in Cyprus: recognition and dating of a Pre-Pottery Neolithic A occupation. *Antiquity* 84: 693–706.

Marchant, R., and H. Hooghiemstra. 2004. Rapid environmental change in African and South American tropics around 4000 years before present: a review. *Earth Science Reviews* 66: 217–60.

Marsh, N. D., and H. Svensmark. 2000. Low cloud properties influenced by cosmic rays. *Physics Review Letters* 85: 5004–7.

Masarik, J., and J. Beer. 1999. Simulation of particle fluxes and cosmogenic nuclide production in the earth's atmosphere. *Journal of Geophysical Research* 104: 12099–111.

Mathers, C., and S. Stoddart, eds. 1994. *Development and decline in the Mediterranean Bronze Age*. Sheffield: J. R. Collis Publications.

Mauquoy, D., B. van Geel, M. Blaauw, and J. van der Plicht. 2002. Evidence from northwest European bogs shows "Little Ice Age" climatic changes driven by variations in solar activity. *The Holocene* 12: 1–6.

Mauquoy, D., B. van Geel, M. Blaauw, A. Speranza, and J. van der Plicht. 2004. Changes in solar activity and Holocene climate shifts derived from ^{14}C wiggle-match dated peat deposits. *The Holocene* 14: 45–52.

McCormac, F. G., A. G. Hogg, P. G. Blackwell, C. E. Buck, T. F. G. Higham, and P. J. Reimer. 2004. SHCal04 southern hemisphere calibration 0–11.0 cal kyr BP. *Radiocarbon* 46: 1087–92.

McCormick, M., P. E. Dutton, and P. A. Mayewski. 2007. Volcanoes and the climate forcing of Carolingian Europe, A.D. 750–950. *Speculum* 82: 865–95.

Meehl, G. A., J. M. Arblaster, K. Matthes, F. Sassi, and H. van Loon. 2009. Amplifying the Pacific climate system response to a small 11-year solar cycle forcing. *Science* 325: 1114–18.

Mitchell, J. M., Jr., C. W. Stockton, and D. M. Meko. 1979. Evidence of a 22-year rhythm of drought in the western United States related to the Hale solar cycle since the 17th century. In *Solar-terrestrial influences on weather and climate*, ed. B. M. McCormac and T. A. Seliga, 125–43. Dordrecht: D. Reidel.

Moros, M., J. T. Andrews, D. D. Eberl, and E. Jansen. 2006. Holocene history of drift ice in the northern North Atlantic: evidence for different spatial and temporal modes. *Paleoceanography* 21.doi:10.1029/2005PA001214.

Morris, I. 1987. *Burial and ancient society: the rise of the Greek city-state.* Cambridge: Cambridge University Press.

———. 1998. Archaeology and Archaic Greek history. In *Archaic Greece,* ed. N. Fisher and H. van Wees, 1–91. London: Duckworth.

Muscarella, O. W. 1995. The Iron Age background to the formation of the Phrygian state. *Bulletin of the American Schools of Oriental Research* 299/300: 91–101.

Neff, U., S. J. Burns, A. Mangini, M. Mudelsee, D. Fleitmann, and A. Matter. 2001. Strong coherence between solar variability and the monsoon in Oman between 9 and 6 kyr ago. *Nature* 411: 290–93.

Peristykh, A. N., and P. E. Damon. 2003. Persistence of the Gleissberg 88-year solar cycle over the last ~12,000 years: evidence from cosmogenic isotopes. *Journal of Geophysical Research* 108 (A1), 1003, doi:10.1029/2002JA009390.

Perry, C. A., and K. J. Hsü. 2000. Geophysical, archaeological, and historical evidence support a solar-output model for climate change. *Proceedings of the National Academy of Sciences of the United States of America* 97: 12433–38.

Plunkett, G. 2006. Tephra-linked peat humification records from Irish ombrotrophic bogs question nature of solar forcing at 850 cal. yr BC. *Journal of Quaternary Science* 21: 9–16.

Pross, J., U. Kotthoff, U. C. Müller, O. Peyron, I. Dormoy, G. Schmiedl, S. Kalaitzidis, and A. M. Smith. 2009. Massive perturbation in terrestrial ecosystems of the Eastern Mediterranean region associated with the 8.2 kyr B.P. climatic event. *Geology* 37: 887–90.

Reid, G. C. 2000. Solar variability and the earth's climate: introduction and overview. *Space Science Reviews* 94: 1–11.

Reimer, P. J. 2001. A new twist in the radiocarbon tale. *Science* 294: 2494–95.

Reimer P. J., M. G. L. Baillie, E. Bard, A. Bayliss, J. W. Beck, C. J. H. Bertrand, P. G. Blackwell, C. E. Buck, G. S. Burr, K. B. Cutler, P. E. Damon, R. L. Edwards, R. G. Fairbanks, M. Friedrich, T. P. Guilderson, A. G. Hogg, K. A. Hughen, B. Kromer, G. McCormac, S. Manning, C. Bronk Ramsey, R. W. Reimer, S. Remmele, J. R. Southon, M. Stuiver, S. Talamo, F. W. Taylor, J. van der Plicht, and C. E. Weyhenmeyer. 2004. IntCal04 terrestrial radiocarbon age calibration, 0-26 cal kyr BP. *Radiocarbon* 46: 1029–58.

Renssen, H., B. van Geel, J. van der Plicht, and M. Magny. 2000. Reduced solar activity as a trigger for the start of the Younger Dryas? *Quaternary International* 68–71: 373–83.

Renssen, H., H. Goosse, and R. Muscheler. 2006. Coupled climate model simulation of Holocene cooling events: oceanic feedback amplifies solar forcing.

Climate of the Past 2: 79–90. http://www.clim-past.net/2/79/2006/cp-2-79-2006.pdf [last accessed 05 April 2010].

Rind, D. 2002. The sun's role in climate variations. *Science* 296: 673–77.

Roberts, A. P. 2008. Geomagnetic excursions: knowns and unknowns. *Geophysical Research Letters* 35: L17307, doi: 10.1029/2008GL034719, 2008.

Rohling, E. J., and H. Pälike. 2005. Centennial-scale climate cooling with a sudden cold event around 8,200 years ago. *Nature* 434: 975–79.

Rosen, A. M. 2007. *Civilizing climate: social responses to climate change in the ancient Near East*. Lanham: Altamira Press.

Sallares, R. 1991. *The ecology of the ancient Greek world*. London: Duckworth.

Scafetta, N., and B. J. West. 2005. Estimated solar contribution to the global surface warming using the ACRIM TSI satellite composite. *Geophysical Research Letters* 32: L18713 doi:10.1029/2005GL023849.

Scott, E. M., and P. J. Reimer. 2009. Calibration introduction. *Radiocarbon* 51: 283–85.

Shindell, D. T., G. A. Schmidt, M. E. Mann, D. Rind, and A. Waple. 2001. Solar forcing of regional climate during the Maunder Minimum. *Science* 294: 2149–52.

Snodgrass, A. M. 1977. *Archaeology and the Rise of the Greek State*. Cambridge: Cambridge University Press.

———. 1980. *Archaic Greece: the age of experiment*. London: J. M. Dent.

———. 1983. Two demographic notes. In *The Greek Renaissance of the eighth century BC: tradition and innovation*, ed. R. Hägg, 167–71. Stockholm: Svenska Institut i Athen.

Solanki, S. K. 2002. Solar variability and climate change: is there a link? *Astronomy & Geophysics* 43: 9–13.

Sternberg, R. S. 1992. Radiocarbon fluctuations and the geomagnetic field. In *Radiocarbon after four decades: an interdisciplinary perspective*, ed. R. E. Taylor, A. Long, and R. S. Kra, 93–116. New York: Springer.

St-Onge, G., J. S. Stoner, and C. Hillaire-Marcel. 2003. Holocene paleomagnetic records from the St. Lawrence estuary, eastern Canada: centennial- to millennial-scale geomagnetic modulation of cosmogenic isotopes. *Earth Planetary Science Letters* 209: 113–30.

Steinhilber, F., J. Beer, and C. Fröhlich. 2009. Total solar irradiance during the Holocene. *Geophysical Research Letters* 36, L19704, doi:10.1029/2009 GL040142.

Stuiver, M., and T. F. Braziunas. 1988. The solar component of the atmospheric ^{14}C record. In *Secular solar and geomagnetic variations in the last 10,000 years*, ed. F. R. Stephenson and A. W. Wolfendale, 245–66. Dordrecht: Kluwer Academic Publishers.

———. 1989. Atmospheric ^{14}C and century scale solar oscillations. *Nature* 338: 405–8.

———. 1993. Sun, ocean, climate and atmospheric $^{14}CO_2$: an evaluation of causal and spectral relationships. *The Holocene* 3: 289–305.

———. 1998. Anthropogenic and solar components of hemispheric ^{14}C. *Geophysical Research Letters* 25: 329–32.

Stuiver M., T. F. Braziunas, B. Becker, and B. Kromer. 1991. Climatic, solar, oceanic, and geomagnetic influences on late-glacial and Holocene atmospheric $^{14}C/^{12}C$ change. *Quaternary Research* 35: 1–24.

Stuiver, M., and H. A. Polach. 1977. Reporting of ^{14}C data. *Radiocarbon* 19: 355–63.

Stuiver, M., and P. D. Quay. 1980. Changes in atmospheric carbon-14 attributed to a variable sun. *Science* 207: 11–19.

Stuiver, M., P. J. Reimer, and T. F. Braziunas. 1998a. High-precision radiocarbon age calibration for terrestrial and marine samples. *Radiocarbon* 40: 1127–51.

Stuiver, M., P. J. Reimer, E. Bard, J. W. Beck, G. S. Burr, K. A. Hughen, B. Kromer, G. McCormac, J. van der Plicht, and M. Spurk. 1998b. IntCal98 radiocarbon age calibration, 24,000–0 cal BP. *Radiocarbon* 40: 1041–83.

Suess, H. E. 1955. Radiocarbon concentration in modern wood. *Science* 122: 415–17.

———. 1965. Secular variations of cosmic ray produced carbon-14 in the atmosphere. *Journal of Geophysical Research* 70: 5937–52.

Tans, P. P., A. F. M. De Jong, and W. G. Mook. 1979. Natural atmospheric ^{14}C variation and the Suess effect. *Nature* 280: 826–28.

Taylor, R. E. 1997. Radiocarbon dating. In *Chronometric dating in archaeology*, ed. R. E. Taylor and M. J. Aitken, 65–96. New York: Plenum Press.

Taylor, R. E., M. Stuiver, and P. J. Reimer. 1996. Development and extension of the calibration of the radiocarbon time scale: archaeological applications. *Quaternary Science Reviews* 15: 655–68.

Thompson, L. G., E. Mosley-Thompson, H. Brecher, M. Davis, B. León, D. Les, P.-N. Lin, T. Mashiotta, and K. Mountain. 2006. Abrupt tropical climate change: past and present. *Proceedings of the National Academy of Sciences of the United States of America* 103: 10536–43.

Trigger, B. G. 1990. Monumental architecture: a thermodynamic explanation of symbolic behavior. *World Archaeology* 22: 119–31.

———. 2003. *Understanding early civilizations: a comparative study*. Cambridge: Cambridge University Press.

Turetsky, M. R., S. W. Manning, and R. K. Wieder. 2004. Dating recent peat deposits. *Wetlands* 24: 324–56.

Usoskin, I. G., K. Mursala, S. Solanki, M. Schüssler, and K. Alanko. 2004. Reconstruction of solar activity for the last millennium using ^{10}Be data. *Astron Astrophys* 413: 745–51.

Usoskin, I. G., M. Schüssler, S. K. Solanki, and K. Mursula. 2005. Solar activity, cosmic rays, and the earth's temperature: a millennium-scale comparison. *Journal of Geophysical Research* 110: A10102 doi: 10.1029/2004JA010946.

van Geel, B., J. Buurman, and H. T. Waterbolk. 1996. Archaeological and palaeoecological indications for an abrupt climate change in the Netherlands and evidence for climatological teleconnections around 2650 BP. *Journal of Quaternary Science* 11: 451–60.

van Geel, B., O. M. Raspopov, H. Renssen, J. van der Plicht, V. A. Dergachev, and H. A. J. Meijer. 1999. The role of solar forcing upon climate change. *Quaternary Science Reviews* 18: 331–38.

van Geel, B., J. van der Plicht, M. R. Kilian, E. R. Klaver, J. H. M. Kouwenberg, H. Renssen, I. Reynaud-Farrera, and H. T. Waterbolk. 1998. The sharp rise of Δ ^{14}C ca. 800 cal BC: possible causes, related climatic teleconnections and the impact on human environments. *Radiocarbon* 40: 535–50.

van Geel, B., N. A. Bokovenko, N. D. Burova, K. V. Chugunov, V. A. Dergachev, V. G. Dirksen, M. Kulkova, A. Nagler, H. Parzinger, J. van der Plicht, S. S. Vasiliev, and G. I. Zaitseva. 2004. Climate change and the expension of the Scythian culture after 850 BC, a hypothesis. *Journal of Archaeological Science* 31: 1735–42.

Vita-Finzi, C. 2008. Fluvial solar signals. *Geological Society, London, Special Publications* 296: 105–15.

von Grafenstein, U., H. Erlenkeuser, J. Muller, J. Jouzel, and S. Johnsen. 1998. The cold event 8200 years ago documented in oxygen isotope records of precipitation in Europe and Greenland. *Climate Dynamics* 14: 73–81.

Walanus, A. 2009. Systematic bias of radiocarbon method. *Radiocarbon* 51: 433–36.

Walker, M. 2005. *Quaternary dating methods.* Chichester: John Wiley & Sons.

Wang, Y., H. Cheng, R. L. Edwards, Y. He, X. Kong, Z. An, J. Wu, M. J. Kelly, C. A. Dykoski, and X. Li. 2005. The Holocene Asian monsoon: links to solar changes and North Atlantic climate. *Science* 308: 854–57.

Wanner, H., J. Beer, J. Bütikofer, T. J. Crowley, U. Cubasch, J. Flückiger, H. Goosse, M. Grosjean, F. Joos, J. O. Kaplan, M. Küttel, S. A. Müller, I. C. Prentice, O. Solomina, T. F. Stocker, P. Tarasov, M. Wagner, and M. Widmann. 2008. Mid- to late Holocene climate change: an overview. *Quaternary Science Reviews* 27: 1791–1828.

Weber, S. L., T. J. Crowley, and G. van der Schrier. 2004. Solar irradiance forcing of centennial climate variability during the Holocene. *Climate Dynamics* 22: 539–53.

Weiss, H., M.-A. Courty, W. Wetterstrom, F. Guichard, L. Senior, R. Meadow, and A. Curnow. 1993. The genesis and collapse of Third Millennium North Mesopotamian civilization. *Science* 261: 995–1004.

Weiss, H. 2000. Beyond the Younger Dryas: collapse as adaptation to abrupt climate change in ancient west Asia and the eastern Mediterranean. In *Environmental disaster and the archaeology of human response,* ed. G. Bawden and R. M. Reycraft, 75–98. Albuquerque: Maxwell Museum of Anthropology.

Weninger, B. 1990. Theoretical radiocarbon discrepancies. In *Thera and the Aegean world III: chronology,* ed. D. A. Hardy and A. C. Renfrew, 216–31. London: The Thera Foundation.

Wenke, R. J. 2009. *The Ancient Egyptian state: the origins of Egyptian culture (c.8000–2000 BC).* Cambridge: Cambridge University Press.

White, W. B., J. Lean, D. R. Cayan, and M. D. Dettinger. 1997. Response of global upper ocean temperature to changing solar irradiance. *Journal of Geophysical Research* 102 (C2): 3255–66.

White, W. B., M. D. Dettinger, and D. R. Cayan. 2003. Sources of global warming of the upper ocean on decadal period scales. *Journal of Geophysical Research* 108 (C8): 3248, doi: 10.1029/2002JC001396.

Whitley, J. 2001. *The archaeology of ancient Greece.* Cambridge: Cambridge University Press.

Wigley, T. M. L., and D. S. Schimel. 2000. *The carbon cycle.* Cambridge: Cambridge University Press.

Zhang, P., H. Cheng, R. L. Edwards, F. Chen, Y. Wang, X. Yang, J. Liu, M. Tan, X. Wang, J. Liu, C. An, Z. Dai, J. Zhou, D. Zhang, J. Jia, L. Jin, and K. R. Johnson. 2008. A test of climate, sun, and culture relationships from an 1810-year Chinese cave record. *Science* 322: 940–42.

3

Animals in Rock Art as Climatic Indicators

PAUL G. BAHN

INTRODUCTION

In any attempt to use animal depictions in rock art for the reconstruction of changes in climate, there are three major obstacles to overcome. The first and most basic of these is the identification of the species represented; the second is the question of whether species' environmental indicators were the same in the past as they are at present; and the third, of course, is the dating of the depictions. None of these three obstacles is ever easy to tackle, especially the last.

IDENTIFICATION OF SPECIES

Even the most basic interpretations of prehistoric rock art—the identification of figures—may well be wrong. In Ice Age cave art, for example, most of the animal figures seem to be easily identifiable—to our twenty-first-century Western eyes—when they are clear and naturalistic. However, there are many figures that are incomplete, stylized, schematic, abbreviated, or perhaps just badly drawn, and that are impossible, therefore, to identify with any degree of certainty.

Much the same applies to every other corpus of rock art but, even where "easily identifiable" figures are concerned, in the absence of the artist(s), we can never be sure that our identifications are correct. In a famous Australian test case, a non-Aboriginal researcher, N. W. G. Macintosh (1977), identified a number of animal images using zoological reasoning, only to learn from an Aboriginal informant that, out of twenty-two images he had been wrong on about fifteen and only superficially right about the other seven! Nevertheless, it is obvious that, for the most part, we must rely on common sense in order to use rock art as a source of information. If it looks like an elephant, and every single person who sees the depiction recognizes it as an elephant,

Figure 1. Crocodile from Libyan Sahara.

then one can be reasonably sure—albeit never completely certain—that it is what the artist intended us to see. This is the best we can hope for (Fig. 1).

ENVIRONMENTAL INDICATORS

The climatic and vegetational preferences of living species are very well known, and one has to make the basic assumption, therefore, that these species had the same preferences in the past, even the remote past, as in the present. On the face of it, this assumption seems reasonable; after all, archaeologists constantly have to compound it with the assumptions that animals' behavior in the past is predictable; and that, human behavior itself is predictable in the past. These are considerable steps to take, but without them archaeology is impossible. Where extinct species are concerned, the problem is somewhat more difficult. A study of the pollen and sediments in which their remains are found, together with the range of other species—especially of microfauna—accompanying them, provides a reasonably solid indicator of the climatic preferences, or more accurately tolerances, of these now vanished creatures. In the case of mammoths and mastodons, we even have the stomach contents of preserved specimens, and the contents of dung from arid sites, to add to the environmental picture (Lister and Bahn 2007).

DATING OF ROCK ART

One of the most intractable problems in the study of rock art is its dating, because at present there is no reliable method of discovering the precise age of the vast majority of it—engravings, petroglyphs, and any inorganic pigments (Bahn 1998; Bahn and Vertut 1997). Instead, one has to date indirectly by style, except in exceptional circumstances such as where the art is masked by datable sediments; or has fallen off and become stratified in datable layers; or is in a position that would have been inaccessible in certain periods (e.g., in high mountains, or on previously drowned coast- or shore-lines). Even where direct dating can be carried out—such as in charcoal drawings from the Ice Age—the uncertainties of the method, involving such problems as minute samples, risks of contamination, etc., mean that the dates obtained may not always be reliable. For most rock art in the world, dating sequences have to be worked out through style, content, and location, and consequently much remains uncertain. For example, the question of whether any rock art in the Sahara can safely be attributed to the Pleistocene period remains unresolved and controversial. Ice Age parietal art, however, is unique in the world of rock art; alongside it we have thousands of pieces of portable art that can be dated either directly or through stratification, and with which its style and content and techniques can be directly compared, usually with very convincing and consistent results.

In Europe, proof of Palaeolithic age can take a number of forms: for example, the depiction of animals that are now extinct (such as the mammoth, which had disappeared from mainland Eurasia by the end of the Pleistocene [Lister and Bahn 2007]), or that were only present during the Ice Age (figures of reindeer in southern France and northern Spain [Fig. 2]), are solid arguments that, in fact, first convinced the nineteenth-century scientific world of the reality of Palaeolithic portable art (Bahn and Vertut 1997).

In the past, attempts were made to date Palaeolithic parietal art by the species depicted. It was assumed that the pictures reflected the faunal assemblage outside, so that the Spanish cave of Las Monedas, for example—which has mostly horses (42%) with some reindeer (13%)—represented a cold phase; whereas the neighboring cave of Las Chimeneas—which has 42% cervids, few horses, and no reindeer—represented an older, warmer phase (González Echegaray 1968, 1974: 39–42). Both sets of these figures, however, seem to be of the same style, and the direct dates they have recently produced point to a far more complex situation.

Whereas Monedas and Chimeneas once seemed to be simple, homogeneous "sanctuaries," matters have become far more complicated for heterogeneous collections of animal figures, such as the 155 in the Spanish cave of El Castillo, which have been assigned to the Aurignacian (27), Gravettian (8), Solutrean (25), Lower Magdalenian (88), and Upper Magdalenian (7) periods. The ten species represented are not particularly indicative of a cold or warm climate; and if the chronological attributions are correct, then there was a massive presence of red deer in the Solutrean, whereas only bison existed in the Upper Magdalenian (González Echegaray 1972).

Figure 2. Engraved reindeer in the Spanish Basque cave of Altxerri.

Clearly, there are many other possible reasons for these differences in depicted species; they should not be taken simply as a tally of what was available outside. Species percentages in depictions rarely correspond to those in animal bones; both assemblages are a conscious selection of what was available. While a picture of a reindeer may prove that the animal was present, *an absence of reindeer pictures does not prove that the animal was no longer around;* and, in any case, parietal and portable art of the same period, and even in the same site, tends to depict different species. Thus, at the French Pyrenean cave of Le Tuc d'Audoubert, there are close analogies between the two art-forms in style and details of execution, and the two are clearly contemporaneous. The portable material has been dated to 12,400 BC. Yet the bison dominates on the walls, and felines and reindeer are present, but there are no fish or anthropomorphs; the portable art, on the other hand, has no felines or reindeer, but it does have fish and anthropomorphs, and it is dominated by horses (Bégouën and Clottes 1985: 43).

REGIONAL VARIATIONS

Assuming we can correctly identify the animals, which, as is shown above, is not always certain, then they can tell us a great deal about the environment. It needs to be borne in mind that this may not necessarily be the local environment; for example, ibex are depicted at the French Ice Age cave of Pair-non-Pair near Bordeaux, but bones of this species have never been found in that region. Similarly, what appears to be an ibis is depicted in the late Ice Age cave art of Church Hole at Creswell Crags, England (Fig. 3) but, once again, bones of this species have not been found in prehistoric Britain so far. In short, the fauna depicted may provide evidence for the local environment in some cases, but in others it may be indicating the environmental conditions of the period in a far wider region.

Some deserts contain depictions of a rich fauna. For example, there are images of camelids all over Chile's Atacama Desert, both in geoglyphs and in rock art (Fig. 4), because throughout late prehistory and history caravans of camelids constantly crossed the desert, from the Andes to the Pacific coast and back. These depictions, therefore, do not really provide any guide at all to changed climatic conditions. Only the rock art of the Sahara—especially the Libyan Messak and the Akâkus—displays graphic (in every sense) evidence of radical shifts in climate and environment through time, as we shall see below.

ICE AGE EUROPE

Some species do not necessarily indicate conditions different from today; they are simply extinct. One such example in Western Europe is the Megaloceros, or extinct giant deer. Others, such as the mammoth, were probably cold-tolerant rather than necessarily indicative of full glacial conditions; but on the

Figure 3. Ibis from Church Hole, Creswell Crags, England.

Figure 4. Camelids in the Atacama Desert.

whole, they indicate cold temperatures, which is why their bones and depictions are not found south of Cantabria (El Castillo) and Asturias (El Pindal) in Spain. This is also why the Ice Age open-air art of southwest Europe, which has survived farther south in Spain and Portugal, only depicts the fauna of horse, aurochs, ibex, and deer that were the staple resources of the period in those regions. No extinct or cold-indicating species are depicted here.

As with the mammoth, the southern limit of Ice Age depictions of reindeer occurs in northern Spain. It is noteworthy that, in parietal art at least, the reindeer is markedly more prominent in this region than in southern France, with beautiful figures in the caves of Tito Bustillo, Las Monedas, and Altxerri. Perhaps the animal was deemed more worthy of depiction in Spain because it was more rare there and thus more of a novelty. In any case, depictions of reindeer and mammoth both provide pretty solid indicators of much colder conditions in Ice Age southern Europe. However, there are distinct regional differences in the animals represented on the walls of the Ice Age caves.

For example, one interesting analysis of the animals represented in the cave art of Spain (Altuna 2002) has shown that out of a total of 2,160 figures, there are 667 red deer (30.9%), 598 horses (27.7%), 328 ibex (13.5%), 251 bison (12%), and 216 aurochs (10.4%). There is thus a clear contrast with the Périgord region, where the horse dominates, followed by fewer bison and mammoth, with deer more rare (reindeer outnumber red deer); and with the Pyrenees, where more than half the figures are bison, followed by the ibex. The Spanish Basque country is a kind of transitional area, with the bison far more abundantly depicted there than in the rest of northern Spain (47.5% of all animals, unlike 11% in Cantabria and only 7.7% in Asturias); the red deer is only 8.9% in the Pais Vasco, but 35.4% in the rest of northern Spain. In the rest of the peninsula, the horse dominates (40.5%), with red deer second (23.4%), followed by ibex (16.7%) and aurochs (14.9%), while the bison is only 1.1%. In the open-air Ice Age art of Foz Côa and Siega Verde, the horse dominates (42.7%), followed by aurochs (20.3%), ibex (18.9%), and deer (17.8%).

In a different study of the fauna depicted in the interior of the peninsula, it has been shown clearly that choice of species reflects cultural rather than environmental factors (Alcolea González and Balbín Behrmann 2003), another important point to bear in mind in any attempt to build precise environmental models on the basis of rock art.

NORTHERN AUSTRALIA

In North Australia, it has proved possible to link rock art to some major changes in coastal environments. According to a study carried out in the Kakadu region by Chaloupka (1984, 1993), as the sea rose at the end of the last Ice Age, it caused changes in the local plants and animals, which in turn produced modifications in technology, all of which seem to be reflected in the region's art. The deduced variations in sea level are themselves important in providing a date for the art. Chaloupka's pre-Estuarine period, broadly coinciding with the height of the last glaciation, depicts non-marine species,

including several that have been interpreted as animals now extinct. In the Estuarine period (starting 6,000 or 7,000 years ago, by which time the post-glacial rise in sea level had ceased), one finds images of new species such as the barramundi (giant perch) and the saltwater crocodile, whose presence can be explained by encroaching seawater that had partially filled the shallow valleys and creeks, creating a salt-marsh environment. Contemporaneously, other species, such as small marsupials, which had once occupied the pre-estuarine plains, now moved further inland and disappeared from the coastal art, as did the boomerang, the weapon used to hunt them. Finally, the Freshwater period (about 1,000 years ago) brought another great environmental change, when freshwater wetlands developed, supporting species of waterfowl and new food plants such as lilies and wild rice, all of which are depicted in the rock art.

THE SAHARA

In the prehistory of the Sahara, aridity seems to have arisen episodically, and very dry periods have been followed by more humid periods that favored human occupation. The desert grew and shrank at different times. There was a major humid phase around 8000 BP, known as a "pluvial." An extremely arid phase followed after 7500 BP and lasted for 1,000 years (but with different durations in different regions). The humid "neolithic" period followed, from about 6500 to 4500 BP. By 2500 BP, the Sahara was more or less as it is now (Gauthier et al. 1996).

During humid phases, the Sahara clearly contained all the big African fauna—elephants, hippos, rhinos, crocodiles—all of which are reproduced in its rock art (see Figs. 1 and 5). Indeed, about one hundred hippo figures are known in the Central Sahara (Gauthier et al. 1996: 61). As mentioned earlier, virtually none of this rock art can be dated directly. Various chronological and cultural sequences have been presented over the years, based largely on occupation dates obtained from archaeological excavations, while pollen analyses help to reconstruct the climate and environment for each period.

The rock art is clearly linked to the different climatic phases. The earlier phases have images of extinct buffalo and aurochs, while the later periods depict the present-day fauna—camels, mouflons, oryx, and gazelles, typical of arid regions. However, as with Ice Age cave art and any other rock art corpus, this is definitely not a bestiary, a catalogue of the species that were around at any particular time. Culture always leads to choices, and some species are underrepresented, while others are totally absent. Moreover, it is now known that the extinct buffalo disappeared quite late (after 4000 BP), and hippo bones have been found as late as 3500–2000 BP, so depictions of these species cannot be taken as reliable markers of early periods (Le Quellec 2006).

Where depictions of domestic bovids and ovicaprids are concerned, however, one can be sure that they postdate the seventh millennium BP, which is when they appear in the archaeological record. Herders seem to have appeared in the Sahara after the mid-Holocene Arid period, around 6000 BP,

Figure 5. Sahara as savanna.

Figure 6. Saharan bovids including a milking scene.

and produced petroglyphs that were devoted above all to their domestic bovids (Fig. 6). As the climate deteriorated during the Post-Neolithic Arid period (starting ca. 5000 BP), cattle herding declined and people turned instead to ovicaprines, which effectively replaced cattle around 4000 BP. Meanwhile, the big African animals were steadily disappearing (Le Quellec 2006).

CONCLUSION

To sum up, there are tremendous uncertainties involved in using rock art's depictions of fauna in any reconstruction of environment or climate. Quite apart from the basic problems of identification of species and dating the images, it is clear from the examples shown above that the range of animals depicted is often dictated more by culture than by ecology, that such depiction groups cannot be taken as accurate bestiaries, and that lack of depictions of a particular species by no means proves its absence. Nevertheless, there are cases—such as Ice Age cave art, and art in northern Australia—where it does appear that valid conclusions can occasionally be reached about environmental change, based on changes in the faunal content of rock art. It is likely that, in the future, as our ability to date different kinds of rock art improves, we will be able to produce far more detailed and accurate chronological sequences that can be related to equally improved environmental sequences based on sedimentological and palynological data. Hence, it is hoped that we shall eventually acquire a far clearer picture of human adaptations to changes in climate.

REFERENCES

Alcolea González, J. J., and R. de Balbín Behrmann. 2003. Témoins du froid. La faune dans l'art rupestre paléolithique de l'intérieur péninsulaire. *L'Anthropologie* 107: 471–500.

Altuna, J. 2002. Los animales representados en el arte rupestre de la Península Ibérica. Frecuencias de los mismos. *Munibe* 54: 21–33.

Bahn, P. G. 1998. *The Cambridge illustrated history of prehistoric art.* Cambridge: Cambridge University Press.

Bahn, P. G., and J. Vertut. 1997. *Journey through the ice age.* London: Weidenfeld Nicolson; Berkeley and Los Angeles: University of California Press.

Bégouën, R., and J. Clottes. 1985. L'art mobilier des Magdaléniens. *Archéologia* 207 (November): 40–49.

Chaloupka, G. 1984. *From palaeoart to casual paintings.* Northern Territory Museum of Arts and Sciences, Darwin. Monograph 1.

———. 1993. *Journey in time.* Chatswood, NSW: Reed.

Gauthier, Y., C. Gauthier, A. Morel, and T. Tillet. 1996. *L'Art du Sahara.* Paris: Le Seuil.

González Echegaray, J. 1968. Sobre la datación de los santuarios paleolíticos. In *Simposio de arte rupestre, Barcelona 1966*, ed. E. Ripoli, 61–65. Barcelona: Diputación provincial de Barcelona.

————. 1972. Notas para el estudio cronológico del arte rupestre de la cueva del Castillo. In *Santander symposium*, ed. M. Almagro Basch and M. A. García Guinea, 409–22. Santander and Madrid: Patronato de las Cuevas prehistóricas de Santander.

————. 1974. *Pinturas y grabados de la cueva de las Chimeneas (Puente Viesgo, Santander)*. Monografías de Arte Rupestre, Arte Paleolítico No. 2. Barcelona: Diputación provincial de Barcelona.

Le Quellec, J.-L. 2006. L'adaptation aux variations climatiques survenues au Sahara central durant l'Holocène. In *Le Sahara et l'homme: un savoir pour un savoir-faire*, ed. M. H. Fantar, 109–29. Actes colloque Douz, December 2003. Tunis: Université de Tunis El Manar.

Lister, A., and P. G. Bahn. 2007. *Mammoths*, 3rd ed. London: Marshall.

Macintosh, N. W. G. 1977. Beswick Creek Cave two decades later. In *Form in indigenous art*, ed P. J. Ucko, 191–97. London: Duckworth.

4

AD 536: The Year Merlin (Supposedly) Died

RICHARD HODGES

Something mysterious and unusual seems to be coming on us from the stars. . . . For things in mid space dominate our sight, and we can see through them only what the rarity of their substance allows . . . for this vast inane, which is spread between earth and heaven as the most tenuous element, allows us to see clearly as long as it is pure, and splashed with the sun's light. But if it is condensed by some sort of mixture, then, as if with a kind of tautened skin, it permits neither the natural colours, nor the heat of the heavenly bodies to penetrate . . . hence it is that, for so long, the rays of the stars have been darkened with an unusual colour.

—Cassiodorus, *Variae*, trans. Barnish (1992), Book XII, letter 25

[D]uring this year a most dread portent took place. For the sun gave forth its light without brightness . . . and it seemed exceedingly like the sun in eclipse, for the beams it shed were not clear.

—Procopius, *History of the Wars*, trans. Dewing (1916), IV, XIV, 5–6

The sun darkened and its eclipse lasted one and a half years, that is, eighteen months. Although the rays were seen two or three hours [a day], they were weak, just as fruit never became ripe and all wine tasted like sour grapes.

—*The Chronicle of Zuqnīn*, trans. A. Harrak (1999)

INTRODUCTION: NEW CLIMATIC CONTEXTS

Archaeologists tend to be unsympathetic to events as agents of change. Where once in Victorian times excavation narratives were driven by catastrophes or violent sacks, nowadays it is process that informs the contemporary discipline. There is something of a paradox here, of course, because the archaeological

*My thanks to Kim Bowes and John Moreland for critical comments on the ideas reviewed here.

strata do not as a rule reflect long series of processes but rather a vertical collection of (long or short) moments when a deposit was made or, for example, a building was erected or altered or abandoned. Rarely are we able to demonstrate that the traces of a deposit or building were formed over a long period. Even new techniques in micro-stratigraphy are revealing a palimpsest of events—resurfacing beaten earth floors, for example—as opposed to their continuous daily use (Matthews 2005).

A further aspect of this curious paradox is that historians of later Roman and early Medieval Europe are drawn to climatic catastrophes and events such as plagues like moths to a flame. The reason is simple. Clearly, the Roman populations throughout much of Western Europe collapsed, and it was perhaps not until the high Middle Ages that the population regained the numbers that had existed in the second or third centuries. Of course, how are we to explain such a dramatic demographic collapse? The written sources, such as Cassiodorus, Procopius, and Zuqnīn cited above, understandably made much of events—catastrophes and plagues. So, in a recent book on the bubonic plague in the AD 530s and AD 540s of the Justinianic period, a group of distinguished historians describe the plague with very little plausible archaeological evidence (Little 2006). Nothing as comprehensive yet exists for climatic events in early European history, though Baillie's *Exodus to Arthur* (1999; see also Gunn 2000) comes close. In one sense this book with its gaudy cover belongs to the many volumes published at the turn of the present millennium, reviewing millennial thinking. In common with the topic of hominid evolution, the subject of much millennial reexamination, there is a sense that late Roman society, beset with social problems, might also have been at the mercy of natural forces (cf. Malik 2000, 368).

Nevertheless, Baillie's millennial book, like his earlier one, *A Slice Through Time* (1995), examines dramatic climatic circumstances including those in Ireland in the 530s, noted with anguish in the Ulster annals (*Annals of the Four Masters*), and now tellingly confirmed by the dendrochronological evidence. In order to explain the tree-ring information evidence, Baillie seeks the intervention of a comet. The comet, so Baillie explains, makes sense of how King Arthur's wizard-advisor and, perhaps, the priest of the "bright countenance and the long reach," foretold his own death in AD 536 (Baillie 1999, 192–93). It is an intriguing conjunction of modern science and ethnohistory (see also Farquharson 1996; Koder 1996; Keys 1999, 18–19; Gunn 2000, 12; McCormick 2003, 21) (Fig. 1). Now this conjunction has found affirmation from some compelling new evidence. With new analysis of the ice cores by L. B. Larsen and his team (2008), there is much more substantive evidence for some extraordinary climatic events in the 530s, culminating intriguingly in AD 536. Of course, no archaeologist is going to be able to shed new light on Merlin. Instead, the issue is—what impact did this event have upon mid-sixth-century European society?

Larsen and his team, building upon a great deal of previous research, contend that "the improvement in ice-core dating and the increasing availability of high quality SO_2/SO_4 measurements from ice cores from both hemispheres allow us to conclude that a tropical volcanic eruption of somewhat larger

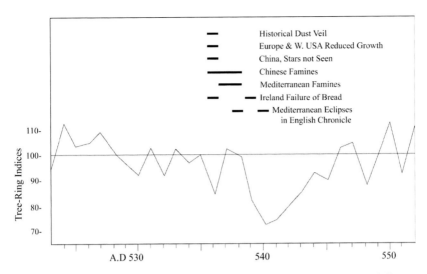

Figure 1. Tree-ring growth shows reduced ring width from 3–8 years following AD 536. Other historical data are indicated above (after Gunn 2000, fig. 1.3).

magnitude than the Tambora eruption most likely caused the 536 dust veil [described by contemporaries]. . . . This removes the apparent mystery that has surrounded the 536 dust veil . . . " (Larsen et al. 2008, 17) (Fig. 2). The dendrochronological evidence, illustrated for example by Baillie (1995; 1999), showed severe summer cooling across wide areas of the Northern Hemisphere, evident in the low growth recorded in a number of long tree-ring chronologies. "It is clear that the European oaks show a strong negative departure in AD 536, dropping on average to 85% of the width in 535. They then recover in 537 and 538 before plunging to less than 75% of their pre-536 width in 540 and 541. The overall chronology does not return to pre-536 values until 546. . . . So, overall, European oaks show an abrupt effect in AD 536, a recovery and reduced growth for some years thereafter" (Baillie 1995, 95–96).

By inference, this indicates "summer cold lasting from 536 to at least 550" (Larsen et al. 2008, 8). This is not the only new scientific data for this brief but critical period. Further support exists in the form of a sharp dip in the CO_2 variations discovered in the Antarctic ice sheet. Here, according to new data from the Law Dome, there was a sharp dip from a high in CO_2 in the first half of the first millennium. William F. Ruddiman, in his book associating human intervention with CO_2 emissions, attributes this extraordinary fall in CO_2 to the coming Little Ice Age (2007, 120–21), but the spectacular change needs more particular explanation (Fig. 3). Larsen and his team, it seems, may have found the clue. But while this might account for the vivid descriptions of veiled light by Cassiodorus in Italy, Procopius in Constantinople, Zuqnīn in Syria, and the Ulster chronicler(s), as well as an account set down several centuries later by the West Saxon chronicler (of the Anglo-Saxon Chronicle) in Wessex, southern England, is it really the basis for an apocalyptic reading of human history in this era?

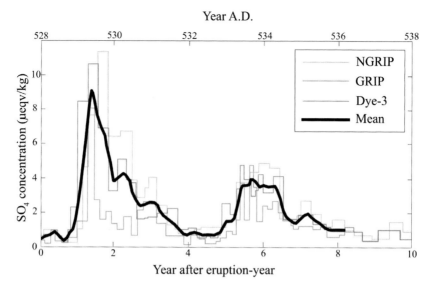

Figure 2. Graph showing SO$_4$ concentrations versus year after eruption-year, after Larsen et al. 2008, fig. 2.

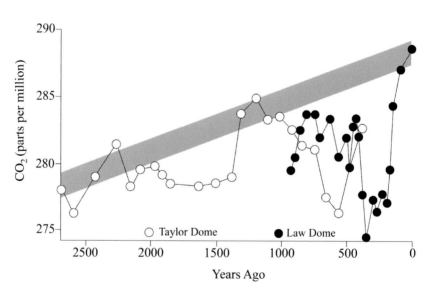

Figure 3. Antarctic ice cores show large oscillations, notably just after the mid first millennium AD (after Ruddiman 2007, fig. 12.1).

Impact on the Mediterranean World?

Most historians would contend that the process of change in the Mediterranean was advanced by the 530s, with roots in the cultural and economic realignments in the seventh century that gave rise to the Islamic revolution in the East and the ascendant "Germanic" communities around the North Sea. Yet the 530s marked an era of significant transformation in the old western Roman provinces. In the East, in Byzantium, the Emperor Justinian was masterminding a revival of his empire using his superior military forces. This involved higher taxes and immense investment in the army and its fortresses. If there is an axiomatic threshold between the Roman and post-Roman worlds, many scholars would ascribe it to precisely this age. Indeed, in AD 536 the Emperor Justinian's finest triumph—a high point of his illustrious reign—was to parade the captured Vandal king and his defeated army through the streets of Constantinople (Young 2000).

Yet there is no indubitable evidence of climatic deterioration. On the contrary, if winters proved to be a little longer or summers were wetter, the evidence essentially eludes detection. However, caught up in the disconcerting conditions created by intense military campaigns, this was an era when longer winters or wetter summers ("We have had . . . a summer without heat . . . the crops have been chilled by north winds . . . the rain is denied" wrote Cassiodorus about the 536 dust veil [Barnish 1992, 12, 25]) would have made a signal difference to peasant economies that rarely exceeded subsistence economic standards and were suffering from taxation to pay for military campaigns (cf. Halstead and O'Shea 1989).

Many if not most peasant households undoubtedly lived off of modest land holdings wherein the margin of a shorter growing season might have had a wholly deleterious impact. We still know too little about the subsistence standards of the Roman peasantry. However, there is ample evidence that by the tenth century a family would have had to cultivate 2–5 hectares of land to provide, at best, marginally sufficient amounts of cereal and other products (cf. Valenti 2008, 153–58, for a Tuscan example). It is not farfetched to envisage that by being dependent upon these small properties families were easily crushed by five or ten years of poor harvests. (See, for example, the modest farm excavated at Pomerance, near Volterra in Tuscany—deserted exactly at this time—described in Camin, Motta, and Terranato 1997.)

Italy, in particular, changed immensely in the period AD 530–50. The long-running struggle between Justinian's armies and the Ostrogoths that coincided with this era is now commonly identified as the catalyst of change. So, while the Byzantine capital of Ravenna and its port at Classe prospered, many smaller inland towns were steadily deserted. More significantly, the Roman rural pattern of lowland living—a legacy of almost a thousand years—suddenly ceased in many regions. Part and parcel of this was the hasty desertion of all categories of sites: from the sprawling bishop's estate centre at San Giusto in northern Apulia (Francovich and Hodges 2003) to the more modest bishopric at upland San Vincenzo al Volturno (Bowes, Francis, and Hodges 2006); from the productive villa in Basilicata at San Giovanni di

Ruoti to the countless farms in the coastal littoral around the old port of Cosa (Francovich and Hodges 2003).

The traditional agrarian base, in other words, was in flux. New excavations show that the peasantry shifted to hilltops such as Miranduolo and Poggibonsi in Tuscany (Francovich and Valenti 1996; Valenti 2004) or, fractionally later, Castel San Donato in Latium (Francovich and Hodges 2003). The precursors of medieval and modern hilltop villages, these places boasted modest crop regimes and livestock economies, and traded only the bare minimum of commodities. The Roman rural economy in much of Italy, in other words, was at or close to an end. Meanwhile, the elite, so it now appears, migrated to certain urban centres, finding security in these places. The import of these upheavals on a war-ravaged society is easily exaggerated. Certainly, the first occupants of Miranduolo (Tuscany) built major terraces in the chestnut woods, established simple but robust sunken-floored huts, and engaged in iron-smelting on a modest yet significant scale (Valenti 2004, 2008). Such conditions were far from aboriginal. Yet the essential economic bases, centred on hilltops as opposed to lowland valleys, had changed forever. Insecurity may explain part of the settlement shift. Equally, given that it occurred ubiquitously in Italy, was this shift also a product of the climatic circumstances brought about by the volcanic eruption in AD 536, leading rural communities to identify new (actually pre-Roman) strategies for survival based upon mixed woodland and lowland use?

The port of Butrint in south-west Albania, with its commercial fortunes tied to the economic fate of Apulia, suddenly went into a marked economic and demographic decline in precisely this era (Hodges, Bowden, and Lako 2004; Hodges 2005). Within the arc of Justinian's reign the town boasted first, the building of a new cathedral-sized basilica and associated baptistery of a certain magnificence, clearly connected to burgeoning commerce with the east Mediterranean, and then the no less notable sudden urban fragmentation with the interment of burials alongside dwellings, breaking an urban code nearly a millennium old. In sum, from the heartlands of the old Roman Empire, the alacrity and nature of settlement transformation has proved surprising to archaeologists and historians alike. Were the 530s a tipping point, abruptly overwhelming a process of late antique transformation that stemmed back as far as the fourth century? Certainly, poor harvests combined with increased levying of taxes (to support the military) would have catapulted most vulnerable communities into high-risk circumstances.

IMPACT UPON THE STRUGGLE BETWEEN THE BRITONS AND ANGLO-SAXONS?

Poor harvests would have had a similarly devastating impact upon southern England. But, unlike Italy, this post-Roman region was essentially underpopulated and split in the era AD 525–50 between sub-Roman communities occupying fortified hilltop sites west of Selwood forest (running north-south from Bath to Dorset) and Anglo-Saxon tribes to the east (Fig. 4). The archaeological

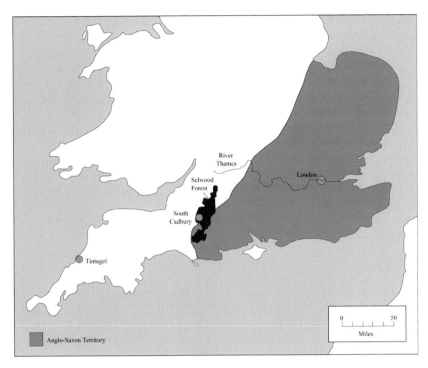

Figure 4. Map showing the location of the Anglo-Saxon tribal areas (shaded), Selwood Forest, South Cadbury Camp, and Tintagel.

evidence from both regions is limited after the withdrawal of the Roman legions in the early fifth century. But in the second and third quarters of the sixth century, the archaeological record is once more available to us. From this time, judging from this record, a profound investment was made by each of these ethnically divided communities to distinguish one from the other. Neither community, as far as we can tell from archaeological surveys, wanted for land or opportunities to expand. But, interestingly, both communities sought extraregional trading partnerships (Campbell 2007) and, famously, the sub-Roman and Anglo-Saxon divisions led to a struggle in which the part played by the legendary Arthur was recalled long after his territories (west of Selwood) had been subsumed into the Anglo-Saxon realm (during the seventh century) (Jones 2000).

The Arthurian world, thanks to excavations of the key site at Tintagel Head, Cornwall, was connected by a tenuous long-distance trade link to the turbulent situation in the western and central Mediterranean (Radford 1956; Barrowman, Batey, and Morris 2007; Campbell 2007, 120–21). The Tintagel excavations first undertaken in the 1930s revealed a sequence of complex structures, mostly made in a traditional Iron Age or Romano-British form. Associated with these was an assemblage of associated ceramics comprising transport amphorae and fine tablewares principally from North Africa and

the Aegean. Traditionally, the imported sherds have been dated to the late fifth or early sixth centuries (cf. Alcock 1971; Radford 1956).

The dates have been significant, causing historians and archaeologists to associate them with the imprecisely dated King Arthur, called Ambrosius by the lugubrious British chronicler Gildas in the mid- to late sixth century (Alcock 1971). Now, new excavations during the 1990s, aided by radiocarbon dates as well as new ceramic studies, point to an occupation focussing upon the second quarter of the sixth century, closer to the time Gildas was actually writing (cf. Barrowman, Batey, and Morris 2007, 313–18; Campbell 2007). More important, this new dating brings the distinctive archaeology of the Arthurian era much closer to the time when the Anglo-Saxon communities were beginning to engage in what Ian Wood christened the Merovingian North Sea (Wood 1993a, 1993b; cf. Campbell 2007, 125–32). Whatever role the rocky central-place at Tintagel played, it dates more accurately an age that extended to fortified hilltops like Cadbury-Camelot on the western edges of Selwood (Alcock 1995), and defines, however modest the trade itself was (Hollinrake 2007, 342), a region that culturally identified itself as "fortress Britain" (Thomas 1998).

Who were the aristocrats of Tintagel and Cadbury-Camelot distinguishing themselves from? The archaeology is quite clear about this (cf. Hines 1997). East of Selwood lay the Anglo-Saxon territories. The political structure of these areas, like western Britain, appears to have been highly fragmented. Judging from the (later) seventh century Tribal Hidage, which depicts about 30 Anglo-Saxon tribal units, it is evident that much of Britain fragmented into numerous tribes with the withdrawal of Roman Imperial government in the fifth century. Indeed, it is possible that as many as 50 to 100 tribal groups existed between c. AD 475 and 550, some coalescing into larger entities from c. AD 550 onward and emerging as kingdoms—the *Angli*, as Pope Gregory the Great defined them—at the end of the sixth century (Hodges 1989, 64–65; Wormald 1994). Sub-Roman Britain, in other words, was a patchwork quilt of Anglo-Saxon and British tribal communities. The archaeology of these communities shows the steady migration westward into the traditional sub-Roman areas— along the Thames valley and from Hampshire, eventually into Dorset.

Around AD 550 the Anglo-Saxon communities began to change significantly. It was then that the Kentish king Ethelberht took a Frankish wife, Bertha, who in turn brought a Christian priest in her entourage. It is a small glimpse of the new cultural pattern evident in the Anglo-Saxon funerary rite that included Frankish prestige goods (swords, dress jewellery, and so forth) as well as a range of east Mediterranean objects ranging from glasses and silverware to Coptic bowls (cf. Harris 2003, 161–88). A Byzantine stamp from London has suggested that Byzantine traders might have been present, not least because the Justinianic chronicler Procopius provides us with an imprecise record of cross-Channel trade at this time (Harris 2003, 176).

Most archaeologists, however, now agree that the traded goods, mostly used in feasting and burial practice, formed part of commercial connections with either northern France (via the Seine) or the Rhineland (cf. Campbell 2007). The goods were given as gifts in a chain of directed relations rather than

traded as cargoes. The pattern of gift-giving, concentrated in southeast and eastern England after c. AD 550, looks to form part of the so-called Merovingian North Sea zone (Wood 1993a). Meanwhile, as is now clear, the western Baltic was in receipt of traded Byzantine gold that was entering this region either from the east Baltic (by way of eastern Europe) or through down-the-line directed gift-giving affecting the tribes of central Europe (Näsman 1998; Randsborg 1998). From the mid-sixth century when the tensions within the patchwork quilt of kingdoms gathered speed, not surprisingly, it seems, competition for resources grew. Most of all, competition to appear Continental grew as well (Harris 2003, 182–87). There were two outcomes: first, a dramatic attempt to conquer the rich farmlands of western Britain, and second, the coalescing of the many small tribes into a few large tribes (Kent, Wessex, Mercia, etc.) to a greater or lesser degree adopting the political apparatus of the Merovingian courts (cf. Hodges 1989, 43–53).

One final word: just as the North Sea communities embarked upon an era of economic interaction that resonated within the chiefly ranks of the Anglo-Saxon communities, so did the Irish Sea communities discover connections in western France (Campbell 2007, 132–39). These connections, post-530s, coincided with the dramatic rise of Irish rural communities (represented by the fortified farmsteads known as raths), by the adoption of new technologies, and, in tandem, the sudden and famed Columbian monastic communities.

Conclusions

So, what (if any) impact did the volcanic veil of AD 536 have on southern England? The answer, notwithstanding the desire to associate Arthur and Merlin with this age, is at best difficult to judge. Undeniably, the Anglo-Saxons had long since embarked upon a steady process of settlement expansion with the likely creation of concomitant tribal coalitions. However, repeated crop failures possibly persuaded the new colonists of the Thames valley to make incursions westward through Selwood forest to assault "fortress Britain." But there is absolutely no archaeological evidence of this climatic event. We may conclude, then, that the recent archaeological evidence defines this period as one of tension, as well as an era in which tribal politics involved the forging of new trading partnerships. Both the "Arthurians" and their enemies, the Anglo-Saxons, were in contact via many intermediaries with Mediterranean sources of commodities, where in the world of Cassiodorus and Procopius entirely different cultural tensions prevailed.

To better understand this event and the full ramifications of its impact north and south of the Alps, we need better archaeological data. First and foremost, we need more examples of peasant dwellings and, in particular, the telltale micro-stratigraphy of their occupation floors (Matthews 2005), where the micro-histories of these people without history may be contained. Less easily obtained might be dendrochronological evidence that, as in Ireland, gives vivid substance to the impact of this "volcanic veil." (So far, no sites in any of these regions have provided preserved timbers that can be "read.")

Finally, we need palynological data: long-term information about the crops and vegetation that might chart any sudden variations.

With new radiocarbon techniques, it is just feasible to build up a micro-history of cropping regimes and equally episodes of sudden change (cf. Rippon, Fyfe, and Brown 2006). Well-dated discoveries—measurements of the impact on ordinary people of this extraordinary event—would also compel historians to scrutinize their texts rigorously, asking, for example, how Byzantine armies procured their provisions in Italy, and how crop failure over multiple years affected tax revenues and indeed the regulation of the market. The sources at face value offer little promise to shed light on such issues, but their silence about such an agrarian catastrophe in itself merits critical examination.

How probable is it that we will discover such archaeological riches for a brief moment in European history? At best, the likelihood is slim, but the compelling character of the new evidence described by Larsen and his team (2008), and the general scepticism of climatic change as a factor in this tumultuous age make the search for such archaeological sites doubly important.

In sum, the decade after AD 536 was an age of signal transformation that affected millions of peasants in western Europe. Did crop failure brought about by the cataclysmic eruption of an unknown volcano possibly located somewhere in Asia contribute to this axiomatic moment in European history? Did the volcanic eruption accentuate the impact of Justinian's imperial ambitions, bringing added torment to the peasantry of this age? Did it contribute to encouraging peasants to quit Roman farming lands and seek safer niches on hilltops? Did it encourage a few adventurous traders visiting Adriatic seaports like Butrint to seek new markets, like those in western Britain? Did it encourage Scandinavian farmers to seek military service in Justinian's armies? And did it disturb the complex tribal divisions in southern England, throwing into relief the resistance of a champion whose history was fashioned and refashioned by European storytellers? Most archaeologists and many historians would be doubtful, but the coincidences are compelling and await conformation in further excavations.

REFERENCES

Alcock, L. 1971. *Arthur's Britain*. Harmondsworth: Penguin Books.

———. 1995. *Cadbury Castle Somerset: the early medieval archaeology*. Cardiff: University of Wales Press.

Baillie, M. G. L. 1995. *A slice through time*. London: Batsford.

———. 1999. *Exodus to Arthur: catastrophic encounters with comets*. London: Batsford.

Barnish, S. J. B., ed. 1992. *The Variae of Magnus Aurelius Cassiodorus Senator*. Liverpool: Liverpool University Press.

Barrowman, R. C., C. E. Batey, and C. D. Morris. 2007. *Excavations at Tintagel Castle, Cornwall, 1990–1999*. London: Society of Antiquaries.

Bowes, K., K. Francis, and R. Hodges, eds. 2006. *San Vincenzo al Volturno 4: from text to territory: excavations and surveys in the monastic terra*. London: British School at Rome.

Camin, L., L. Motta, and N. Terranato. 1997. Pomerance: un sito rurale nel territorio di Volterra. *Archeologia Medievale* 24: 109–16.

Campbell, E. 2007. *Continental and Mediterranean imports to Atlantic Britain and Ireland, AD 400–800*. York: Council for British Archaeology.

Dewing, H. B., trans. 1916. *Procopius, history of the wars: books III and IV*. Loeb Classical Library, London: Heinemann.

Farquharson, P. 1996. Byzantium, planet earth and the solar system. In *The sixth century: end or beginning?*, ed. P. Allen and E. Jeffreys, 263–69. Brisbane: Australian Association for Byzantine Studies.

Francovich, R., and R. Hodges. 2003. *Villa to village*. London: Duckworth.

Francovich, R., and M. Valenti. 1996. The Poggibonsi excavations and the early medieval timber building in europe. In *Archaeology and history of the Middle Ages*, ed. G. P. Brogiolo, S. Gelichi, R. Francovich, R. Hodges, and H. Steuer, 135–49. Forlì: XIII International Congress of Prehistoric and Protohistoric Sciences, Italy.

Gunn, J. D. 2000. AD 536 and its 300 year aftermath. In *The years without summer: tracing AD 536 and its aftermath*, ed. Joel D. Gunn, 5–20. Oxford: Archaeopress.

Halstead, P., and J. O'Shea, eds. 1989. *Bad year economics: cultural responses to risk and uncertainty*. Cambridge: Cambridge University Press.

Harrak, A. 1999. trans. *The chronicle of Zuqnīn*. Toronto: Pontifical Institute of Medieval Studies.

Harris, A. 2003. *Byzantium, Britain and the West*. Stroud: Tempus.

Hines, J., ed. 1997. *The Anglo-Saxons from migration period to the eighth century: an ethnographic perspective*. Woodbridge: Boydell and Brewer.

Hodges, R. 1989. *The Anglo-Saxon achievement*. London: Duckworth.

———. 2005. *Eternal Butrint: A UNESCO World Heritage Site in Albania*. London: Periplus Publishing.

Hodges, R., W. Bowden, and K. Lako, eds. 2004. *Byzantine Butrint: excavations and surveys, 1994–99*. Oxford: Oxbow Books.

Hollinrake, N. 2007. Dark Age traffic on the Bristol Channel, UK: a hypothesis. *International Journal of Nautical Archaeology* 36: 336–43.

Jones, E. 2000. Climate, archaeology, history, and the Arthurian tradition. In *The years without summer: tracing AD 536 and its aftermath*, ed. J. D. Gunn, 25–34. Oxford: Archaeopress.

Keys, D. 1999. *Catastrophe: an investigation into the origins of the modern world*. New York: Ballantine Books.

Koder, J. 1996. Climatic change in the fifth and sixth centuries? In *The sixth century: end or beginning?* ed. P. Allen and E. Jeffrey, 270–86. Brisbane: Australian Association for Byzantine Studies.

Larsen, L. B., B. M. Vinther, K. R. Briffa, T. M. Melvin, H. B. Clausen, P. D. Jones, M.-L Siggard-Andersen, C. U. Hammer, M. Eronen, H. Grudd, B. E. Gunnarson, R. M. Hantemirov, M. M. Naurzbaev, and K. Nicolussi. 2008. New ice core evidence for a volcanic cause of the AD 536 dust veil. *Geophysical Research Letters* 35: L04708.

Little, L., ed. 2006. *Plague and the end of antiquity: the pandemic of 541–750*. Cambridge: Cambridge University Press.

Malik, K. 2000. *Man, beast, and zombie*. London: Weidenfeld & Nicholson.

Matthews, W. 2005. Microstratigraphy and micromorphology: contributions to interpretation of the Neolithic settlement of Çatalhöyük, Turkey. In *Fertile ground: papers in honour of Susan Limbrey*, ed. D. N. Smith, M. B. Brickley, and W. Smith, 108–14. Oxford: Oxbow Books.

McCormick, M. 2003. Rats, communications, and plague: toward an ecological history. *Journal of Interdisciplinary History* 34: 1–25.

Näsman, U. 1998. The Justinianic era of South Scandinavia: an archaeological view. In *The sixth century*, ed. R. Hodges and W. Bowden, 255–79. Leiden: E. J. Brill.

Radford, R. 1956. The imported pottery found at Tintagel. In *Dark-age Britain*, ed. D. B. Harden, 59–70. London: Methuen.

Randsborg, K. 1998. The migration period: model history and treasure. In *The sixth century*, ed. R. Hodges and W. Bowden, 61–88. Leiden: E. J. Brill.

Rippon, S. J., R. M. Fyfe, and A. G. Brown. 2006. Beyond villages and open fields. *Medieval Archaeology* 50: 31–70.

Ruddiman, W. F. 2007. *Plows, plagues and petroleum*. Princeton: Princeton University Press.

Thomas, C. 1998. *Christian Celts: messages and images*. Stroud: Tempus.

Valenti, M. 2004. *L'insediamento a Altomedievale nelle Campagne Toscane. Paesaggi. Popolamento e villaggi tra VI e X secolo*. Florence: Insegna del Giglio.

———, ed. 2008. *Miranduolo in alta val di Merse (Chiusdino—SI)*. Florence: Insegna del Giglio.

Young, B. 2000. Climate and crisis in sixth-century Italy and Gaul. In *The years without summer: tracing AD 536 and its aftermath*, ed. J. D. Gunn, 35–44. Oxford: Archaeopress.

Wood, I. 1993a. *The Merovingian kingdoms*. London: Longman.

———. 1993b. *The Merovingian North Sea*. Alingås: Viktoria Bökforlag.

Wormald, P. 1994. Engla Lond. *Journal of Historical Sociology* 7: 1–24.

5

Did a Bolide Impact Trigger the Younger Dryas and Wipe Out American Megafauna?

STUART J. FIEDEL

A COMET IMPACT AT 12,900 CAL BP?

In May 2007, at the American Geophysical Union meeting in Acapulco, Richard Firestone and his colleagues announced evidence of a bolide impact in North America at 12,900 cal BP.* This presentation has been followed by a flurry of articles and papers (Firestone et al. 2007a, 2007b, 2008; Kennett et al. 2008a, 2008b, 2009, for example).

The current impact hypothesis can be traced back, through its various iterations, to 1998, when William Topping observed tiny pits on the surfaces of flakes found at the Gainey Paleoindian site in Michigan (Topping 1999). He attributed those pits to bombardment by micrometeorites. These particles were said to have been released by a solar flare or supernova shock wave that also was hypothesized to have reset the radiocarbon clock somehow, so that the Clovis culture was really 40,000—not 13,000—years old (Firestone and Topping 2001). Topping is no longer part of the team that is currently pursuing research on the supposed impact, and the disruption of the radiocarbon decay process (effectively refuted by Southon and Taylor 2002) is no longer part of the model.

Based upon recent oral presentations and publications by Firestone, Allen West, Kennett, and their colleagues (e.g., Kennett et al. 2008a, 2008b, 2009), the team now envisions the impactor as either a comet that broke into fragments before it hit the Earth's atmosphere or a swarm of comets or carbonaceous chondrites. The fragments, or swarm, created a series of airbursts comparable to the one that is believed to have hit Tunguska, Siberia, in 1908. An "intense radiation flux" (Kennett et al. 2008a, 2543) ignited continental-scale firestorms and sent high-temperature shock waves across North

*Throughout this discussion, BP will be used for radiocarbon years before present (conventionally set at AD 1950) and cal BP for calendrical years before present. Technically, these are not calibrated ages, because the tree-ring sequence used for calibration does not yet extend earlier than 12,450 cal BP.

America. Effects of the impact are also reported from several sites in western Europe (e.g., the charcoal-rich Usselo soils of the Netherlands) (Firestone et al. 2007a). Various markers of the impact are reported to have been recognized at some thirty sites. They include magnetic iron spherules, carbon spherules, fullerenes, helium 3, iridium, and nanodiamonds. In the Greenland ice cores at 12,900 cal BP, there are peaks in ammonium, nitrate, nitrite, formite, oxylate, and acetate, indicative of the largest and most prolonged interval of biomass burning within the last 110,000 years (West and Kennett 2008).

No crater has been found on the Earth's surface to mark the impact site, and other expected markers of such an event—shocked minerals, breccias, tektites—have not been reported (Kennett et al. 2009). Nevertheless, one subsidiary hypothesis improbably attributes the formation of the Carolina "bays" (ovoid, ponded depressions of the southeastern U.S. coastal plain) to the same event (Kobres et al. 2007). Specialists in that region's geology universally reject such a late, meteoritic origin for these features (Ivester et al. 2004). Another group, including Firestone, has suggested that four "deep holes" under the Great Lakes, and a proposed 1-km diameter impact crater within Lake Ontario, may mark a meteorite impact, and Firestone continues to note the similarity in elemental composition of some microspherules to lunar basalt (Firestone et al. 2008). Furthermore, West and archaeologist Ken Tankersley have suggested that the impactor might have hit the recently discovered diamond region of western Canada, thereby spraying diamonds across the midwestern United States (Hoffman 2008).

Firestone, West, and their colleagues have also recently connected supposed micrometeorites, propelled at high speed into Alaskan mammoth tusks and a Siberian bison skull, with the bolide (Firestone and West 2005; Firestone et al. 2007c). They must concede the chronological difficulty posed by the date of the tusks—ca. 33,000 BP (ca. 37,500 cal BP)—and the skull (ca. 26,000 BP, ca. 31,000 cal BP). Only under the premises of the now-discarded Topping variant of the impact theory (1999), with its disrupted radiocarbon clock, would any association of these finds with the proposed 12,900 cal BP event be even remotely credible. In fact, Firestone and West appear to retain a modified version of Topping's idea, suggesting that the ostensibly much later age of the skull "is consistent with exposure of the bison to an enriched source of radiocarbon following the impact" (http://ie.lbl.gov/Mammoth/Impact.html). Despite the supposed spray of extraterrestrial pellets, there was no extinction event at ca. 33,000 BP; indeed, the bison skull is reported to exhibit bone regrowth over the impacted microparticles, indicating that this animal survived the suggested extraterrestrial assault. With so many deviant and incompatible variants of the model to choose from, it is difficult for critics to pin down the central aspects of the protean impact hypothesis for scientific assessment and refutation (Pinter and Ishman 2008).

In what seems to be the current consensual version of the hypothesis, the comet exploded above the Laurentide ice sheet in eastern Canada (Firestone et al. 2007a). This location would account for the absence of an impact crater. Intense heat from the blast would have melted much of the glacier. The meltwater poured into the North Atlantic, disrupting thermohaline circula-

tion and triggering an abrupt drop in atmospheric temperature in Greenland, Europe, and eastern North America. This marked the onset of the Younger Dryas, a relapse into nearly glacial cold climate that persisted until 11,600 cal BP. The huge wildfires ignited by the blast either directly fried to a crisp North American megafauna or carbonized the vegetation that they needed to subsist (Kennett et al. 2008a). Burned off or starved out, numerous species of megafauna abruptly became extinct. The impact is also supposed to have had a disastrous effect on the human population of North America, causing the rapid demise of the Paleoindian Clovis culture.

THE MYSTERIOUS ONSET OF THE YOUNGER DRYAS

For all its ambiguities and inconsistencies, the new bolide theory does provide a testable explanation for the still-mysterious onset of the Younger Dryas cold interval at ca. 12,900 cal BP. This cold period was first recognized in Europe. Danish paleobotanists observed in 1901 that, as the last Pleistocene ice sheets retreated, open tundra and closed forest vegetation had spread alternately across the landscape. The intervals of tundra were called "Dryas" after a flower of the rose family, *Dryas octopetala* (mountain avens). Preserved macrofossils of this plant and its pollen were very common in deposits from these cold periods, known as the Oldest, Older, and Younger Dryas. After 1950, radiocarbon dates fixed the dates of these periods as about 15,000–12,500 BP (Oldest), 12,000–11,800 BP (Older), and 11,000–10,000 BP (Younger Dryas). These were not the true ages, however. Beginning in the early 1990s, ice cores drilled deep into the Greenland ice cap yielded evidence of widespread cold episodes (stadials) corresponding to these European paleobotanical periods, which could be dated by means of layer-by-layer counting of the annually laid ice deposits. The results were startling in two respects: (1) the real ages were about 2,000 years older than the radiocarbon ages, and (2) the transitions from cold to warm, and back again, were very rapid. Each new study seems to further reduce the duration of these transitions, from a few decades to just a few years. The end of the Oldest Dryas (Greenland Stadial 2) and the rapid onset of the warmer Bølling (or Greenland Insterstadial [GI]-1e) are now dated to 14,650 cal BP. Warm temperatures persisted but fell very gradually over the next 2,000 years (the Bølling-Allerød), interrupted by a few brief cold intervals: the Older Dryas (GI-1d) at about 14,100–13,900 cal BP (separating the Bølling from the Allerød [GI-1abc]) and the Intra-Allerød Cold Period, around 13,200–13,100 cal BP. After a short unnamed warm period (GI-1a), the Younger Dryas (GS-1) began at 12,950, 12,850, or 12,650 cal BP (according to different counts of varves and ice layers). This cold period lasted for about 1,300—or perhaps only 1,100—years; it ended at 11,590 cal BP according to Central European tree rings (Kromer et al. 2004), or 11,653 cal BP (with BP fixed at AD 1950) according to the NGRIP core (Rasmussen et al. 2006; Steffensen et al. 2008).

Brauer et al. (2008) have recently asserted that the Younger Dryas began in central Europe precisely in the winter of 12,679 cal BP. This date is based upon

evidence, from varves (annually layered lake-bottom sediments), of stronger westerly winds and storms roiling the waters of the Meerfelder Maar, a crater lake in Germany. The winds had been intermittently strong since 12,710 cal BP, but the permanent shift occurred in 12,679 cal BP. Muscheler et al. (2008) and Southon (2007) contend (based upon correlation of varying amounts of radioactive beryllium in ice cores with radiocarbon fluctuations) that the recent end of the German floating tree ring sequence lies at 12,500 cal BP, and that the rings record a Younger Dryas (YD) onset signal at about 12,630 cal BP (not 12,900, as previously assumed [Kromer et al. 2004]).

Remarkably precise though they may appear, these dates may not be quite correct. A half-dozen finely laminated environmental records in the Northern Hemisphere provide year-by-year counts of Late Glacial climatic periods: three ice cores in Greenland (GRIP, GISP2, and NGRIP), sediments in the Cariaco Basin off the coast of Venezuela (Lea et al. 2003; Hughen et al. 2000, 2004), and several European lake beds (Meerfelder Maar in Germany, lakes Gosciaz and Perespilno in Poland [Goslar et al. 1995, 2000a, 2000b]). Several other European lake beds are also cited occasionally (e.g., Björck et al. 1996); data from corals found near Tahiti and Barbados (Fairbanks et al. 2005), and stalagmites in southern Chinese caves (Wang et al. 2001) also are relevant. Unfortunately, these sequences are not in complete agreement (Southon 2002, 2007). The discrepancies may be attributable to two factors: (1) counting mistakes, whether caused by human error, instrument problems, or erratic deposition; and (2) reliance on different aspects of the environment, atmosphere, and chemistry to define the events of interest.

The inability of climate scientists to reconcile their estimates for the onset date of the Younger Dryas stands in puzzling contrast with their virtual unanimity on the date of its abrupt end, which is also when the Holocene began. The German and Swiss trees responded to the sudden warming by adding wide growth rings at 11,590 cal BP (with "present" defined as AD 1950)—that is, 11,640 years before AD 2000 (Kromer et al. 2004). In the Cariaco varved sediments, the date is nearly identical—11,580 cal BP. In the GRIP core, the date is 11,550±70, although the GISP2 date is 11,640 (but with the latter's possible counting error of about 250 years, these dates obviously overlap). The most recently published ice-core results (which revise and supersede the GRIP chronology) come from the NGRIP (North Greenland Ice Core Project) (Rasmussen et al. 2006; Steffensen et al. 2008). In this core, oxygen-isotope ratios (a proxy measure of local atmospheric temperature) show that the air over central Greenland first warmed at 11,711±12 cal BP and kept getting warmer until it stabilized, marking the start of the Holocene at 11,651±13 cal BP (this chronology uses a "present" of AD 2000; these dates are 11,661 and 11,601 cal BP, respectively, using AD 1950 as the reckoning date). The NGRIP core also has very high-resolution data for variations in deuterium (mainly an indicator of the sea-surface temperature [SST] wherever the water vapor that fell as snow over Greenland originated), dust and calcium (jointly indicating conditions in Central Asian deserts, the source of the Greenland dust), and sodium (an index of the extent of sea ice surrounding Greenland). The relative timing and tempo of changes in these indices provide clues about the causes of the

Younger Dryas and other abrupt climate changes of the Late Glacial period (beginning about 18,000 cal BP).

In the NGRIP core, the onset of the Younger Dryas is a much longer process than its termination. If we define the onset as the initial drop in Greenland temperature (the oxygen indicator), it occurs at 12,875±59 cal BP (with "present" defined as AD 1950). However, the temperature falls to its minimum 213 years later, at 12,662±74 cal BP. Defined by lowered SST in the nearby North Atlantic waters (deuterium), the onset is dated to 12,847±3 cal BP; within a year, a minimum and persistently low SST had been reached.

In the Cariaco varves, the final warm peak within the GI-1a occurs at 13,007 cal BP (Lea et al. 2003). As defined by oxygen isotopes, the same peak shows up at 13,043 in GISP2 and at 13,012 in the Hulu Cave speleothem in southern China (Wang et al. 2001), but at 12,870 cal BP in GRIP and NGRIP, and even later in German lake varves. This discrepancy must represent a counting error, not an atmospheric lag effect. In Cariaco, between 12,953 and 12,900 cal BP, the inferred sea-surface temperature dropped from 27.4 to 26 degrees Celsius. It fell again by 2.3 degrees (26.4 to 24.1) between 12,848 and 12,800 cal BP, and reached its coldest state (22.8 degrees) at 12,640 cal BP. The Cariaco research team reckons the duration of the Younger Dryas as at least 1,235 years, if they count only from minimum to minimum temperatures, or 1,335 years if they count to and from the mid-points of the transitional periods at either end (Hughen et al. 2004). In Hulu Cave, the midpoint of the initial YD cooling is at 12,823±60, and the end of the YD is at 11,473±100 cal BP; the length of the period is calculated as 1,350 years, plus or minus 120 (Wang et al. 2001).

The Cariaco chronology agrees very closely with that developed from an ice core by the GISP2 team (Alley et al. 1993). They dated the Younger Dryas onset to 12,940±260 cal BP. However, the date for the same event in the GRIP ice core chronology is about 12,650 cal BP (Johnsen et al. 1992, 2001). The varve-based date from Meerfelder Maar in Germany is roughly the same as the GRIP date—12,679 cal BP. In this case, all the records are relevant to what seems to be an identical climatic phenomenon—a dramatic change in air circulation and winds affected both Cariaco and Meerfelder Maar—so that an explanation of the dating discrepancy must be sought elsewhere. The GISP2 and GRIP coring sites were located only thirty kilometers apart in central Greenland, so they must have captured layers of ice from the same isochronous events. The more recently obtained NGRIP core was located a little farther north, but it too should be completely synchronous with the other Greenland cores.

The varves of Lake Gosciaz in Poland, like the German varves, indicate a much shorter span (about 1,150 years) for the Younger Dryas than the ca. 1,300 years seen in the Cariaco and GISP2 records (Goslar et al. 2000a, 2000b). Goslar et al. (2000b) express their bewilderment: "The reasons that could produce a too-long CB [Cariaco Basin] chronology, are difficult to imagine." They also reveal why the Central European varves might underestimate the duration of the YD: "The accuracy of the varve chronology is limited by varves that appear unclear in all analyzed cores" (Goslar et al. 2000b, 336). It is probably easier to miss a few indistinct or eroded varves or ice layers here and there

than to artificially elongate a sequence by adding multiple imagined layers. Therefore, I suspect that the GISP2 and Cariaco counts eventually will prove to be closer to reality than those of GRIP and the European varves, and that the Younger Dryas lasted about 1,230–1,330 years. As everyone agrees that it ended at 11,600 or 11,650 cal BP, that puts the beginning at about 12,980–12,830 cal BP (NGRIP dates the onset at about 12,900–12,850).

Leaving aside for the moment the recently proposed comet impact theory, three basic causes or "forcing" factors have been invoked to explain why the Younger Dryas happened: ocean circulation (Broecker et al. 1985, 1989), atmospheric circulation (Steffensen et al. 2008), and solar radiation (Renssen et al. 2000; Clement et al. 2001). The prevailing ocean-based theory is quite similar to the sequence depicted in the Hollywood film, *The Day After Tomorrow*. In the movie, a modern-day ice age was provoked in a matter of hours by global warming, which resulted in icecap melting and then a rapid shutdown of warm-water circulation in the North Atlantic. The thermohaline circulation (or MOC, meridional overturning circulation) shutdown model (Broecker et al. 1985, 1989; Broecker 1991, 1997) is very similar, albeit with a somewhat less compressed time frame. According to this model, a great gush of cold, fresh water derived from the melting Laurentide ice sheet in eastern Canada swept across the surface of the North Atlantic. It prevented warm, salty water from the southern ocean, which flowed deep below the surface (the Gulf Stream), from rising to the surface. The normal overturning of the ocean water stopped. As a consequence, the atmosphere over the ocean, which would normally have been warmed, remained cold and so, in consequence, did the air over Europe and North America.

This scenario also provides a plausible explanation for a peculiar property of radiocarbon dates from the beginning of the Younger Dryas: within a century of real time (12,900–12,800 cal BP or perhaps 12,760–12,660 [Hua et al. 2009]), dates drop from about 11,000 to 10,600 BP. This "cliff," followed by a 1700-year "plateau" of almost unchanging radiocarbon dates, is manifest in German and Swiss tree rings (Kromer et al. 2004); the same event, presumably synchronous with the European record, is seen in stratified sediments in Alaska (Hajdas et al. 1998) and Argentina (Hajdas et al. 2003). When the MOC stopped, less carbon dioxide moved from the ocean surface zone into deeper water (Goslar et al. 1995; but see Marchal et al. 1999). Radioactive carbon isotopes, produced in the upper atmosphere by cosmic rays, represent just a tiny fraction of the carbon atoms in CO_2 molecules. But, as the carbon dioxide backed up over the ocean, so did the ^{14}C. Thus, any living organisms taking in air at that time would have contained a greater proportion of ^{14}C than in normal circumstances, and this makes their date of death (the beginning of radioactive decay in their tissues without replacement) seem younger than the actual date.

An alternative explanation of the radiocarbon "cliff" would be that the sun weakened; this would have been accompanied by weakening of the Earth's magnetic field and greater penetration by cosmic radiation. The increased bombardment by cosmic rays would lead to more ^{14}C in the atmosphere, regardless of ocean-surface temperatures. The reduced insolation alone might

also account for the colder temperatures of the Younger Dryas (Renssen et al. 2000). Similar, albeit smaller-scale, ^{14}C cliff-plateau events are associated with cold intervals during the Holocene, marked by iceberg-rafted sand and silt grains in North Atlantic sediments ("Bond events") and apparently driven by solar fluctuations (Bond et al. 2001).

The ice sheets in midlatitudes had begun to melt and recede about 17,400 cal BP (Schaefer et al. 2006). Melting must have accelerated about 14,700 cal BP (the abrupt start of the Bølling warm period), as indicated by a sharp rise in global sea level, called Meltwater Pulse IA, not long afterward (dates varying between 14,700 and 14,000 cal BP have been proposed for the rise [Weaver et al. 2003; Stanford et al. 2006]). Why would continued melting of the ice have caused such a sharp climatic reaction as the YD cooling only at the relatively late date of about 12,900 cal BP (when there was no significant sea level rise)? Proponents of the ocean-forcing theory suggest that this is when meltwater from the Laurentide ice sheet (in eastern Canada) was re-routed eastward, via the St. Lawrence River, into the Atlantic. Previously, the water had flowed into the Mississippi drainage and emptied into the Gulf of Mexico. Cessation of this southern flow has recently been dated to 10,970±40 BP (Williams et al. 2008).

Seemingly crucial for this model is the behavior of glacial Lake Agassiz. This enormous lake formed initially as the ice retreated from northern Minnesota and North Dakota. Ultimately, the lake covered some 365,000 square miles, including portions of Manitoba, Ontario, and Saskatchewan. In Manitoba it was about 600 feet deep. The lake had three outlets through which its water could escape: one in the northwest, draining through the Mackenzie River toward the Arctic Ocean; one in the south, draining through the Minnesota River into the Mississippi, and thence to the Gulf of Mexico; and an eastern outlet draining into the Great Lakes and then through the St. Lawrence River into the North Atlantic. A concentrated research effort was initiated several years ago to establish definitive dates for the presumably catastrophic drainage of the lake waters through the eastern and northwestern outlets. The preliminary results are disappointing; in neither area is there evidence for a lake-water discharge prior to the *end* of the Younger Dryas (Lowell et al. 2005). The hypothesized cause of the switch-off of thermohaline circulation now seems more than a thousand years later than its supposed effect. The discouragement of the researchers is palpable in their conclusion: "These investigations indicate that the geological understanding of past abrupt climate changes is only preliminary. This does not bode well for predicting future, abrupt climate changes" (Lowell et al. 2005, 372).

However, scientists have not yet abandoned the meltwater-trigger model. In another recent study, Carlson et al. (2007) analyzed chemical ratios in the tests (shells) of planktonic foraminifera that accumulated in sediments at the mouth of the St. Lawrence estuary. They conclude that freshwater flow increased sharply around 12,900 cal BP, sufficient to cause the Younger Dryas by upsetting the meridional overturning circulation. Strontium-isotope ratios suggest that the source of the water was in runoff from exposed bedrock surfaces in the western Canadian Shield. However, these inferences cannot be

reconciled with the absence of an eastern outlet of appropriate age in Lake Agassiz. Accordingly, Carlson and his colleagues have been harshly criticized by Peltier et al. (2008), who note that a previous study of dinoflagellates in sediments at the mouth of the St. Lawrence (de Vernal et al. 1996) found no evidence of a freshwater flux at 12,900 cal BP.

Tarasov and Peltier (2005, 2006) still advocate a meltwater trigger for the Younger Dryas, but they propose a very different pathway—into the Arctic Ocean. The ocean circulation system would have been very sensitive to an input in that area. They are forced to acknowledge, however, that field research in the area of the northwestern outlet of Lake Agassiz, as in the east, has also failed to identify a channel dating to 12,900 cal BP (but see now Murton et al. 2010). Accordingly, they minimize the role of Lake Agassiz outflow and focus instead on the melting of the Keewatin Dome, a great mound of ice in the northwestern section of the Laurentide ice sheet. Tarasov and Peltier are rather vague about the mechanism by which the ice and water from this area made its way into the ocean. As for the reason that the dome should have melted just at 12,900 and not a thousand years earlier, they offer as "one possibility . . . the displacement or switching of a 'super-chinook' arising from the impact of the large Cordilleran Ice Sheet on a displaced jet stream" (Tarasov and Peltier 2006, 685). A chinook is a dry wind that occasionally creates anomalous warm weather on the lee side of mountain ranges on the Northwest Coast.

Tarasov and Peltier's invocation of a "displaced jet stream" seems to create a "chicken and egg" problem. Remember that the first signs of the Younger Dryas onset are markers of stronger western winds in central Europe, stronger trade winds in the Caribbean, and a change of the source area from which water vapor reached Greenland, due to "a reorganization of atmospheric circulation from one year to the next" (Steffensen et al. 2008, 681). If major changes in wind patterns caused the ice-melting that shut off thermohaline circulation, obviously we cannot regard the latter as the trigger for the rapid atmospheric transition. In summary, the cause of the Younger Dryas onset is not well understood at present: "Neither the magnitude of such shifts nor their abruptness is currently captured by state-of-the-art climate models" (Steffensen et al. 2008, 683).

Until very recently, those who attributed the demise of the American megafauna to the stresses of changing climate and vegetation argued that the cause of extinction was the gradual transition at ca. 10,000 BP from the Pleistocene to the warmer but more seasonally variable climate of the Holocene (e.g., Graham and Lundelius 1984). However, confronted with the new climate data and a growing corpus of radiocarbon dates for late-surviving fauna, they now must shift the blame to the abrupt and catastrophic onset of the Younger Dryas cold period (Graham 1998; Graham et al. 1997, 2002). For the bolide impact hypothesis, the colder climate in the wake of the YD onset is immaterial; extinction would have been virtually instantaneous due mainly to extreme heat. It is, therefore, most important to observe that the NGRIP record, which allows an objective comparison of the rapidity of deglacial climate changes, shows that the warming trends at the start of the Bølling (14,650 cal BP) and at the end of

the Younger Dryas (11,650 cal BP) were much more abrupt than the cooling at the YD onset (12,850 cal BP). Yet, in neither case did very rapid warming, in just a few years, have any observed dire effect on big mammals. No one has postulated an extraterrestrial trigger for these sudden climate changes.

Another aspect of the timing of these abrupt climate changes offers an important clue to the cause of the megafaunal collapse. For a while, climate scientists debated whether there was a simultaneous Younger Dryas cold spell in the southern hemisphere. Three areas have figured most prominently in these discussions: Antarctica, Patagonia, and New Zealand. As in Greenland, several long cores (Vostok, Byrd, Taylor Dome, Dome C) have been drilled deep into the Antarctic ice cap, revealing a record of glacial and interglacial episodes extending back as early as 800,000 cal BP (EPICA community members 2004). Unlike Greenland, the Antarctic ice lacks clearly defined annual layers, but the amounts of atmospheric gases such as methane, carbon dioxide, and oxygen trapped in bubbles within the ice at different depths permit fairly precise correlation with the Greenland cores and with marine sediments.

However, the climate sequences derived from the several cores have not been aligned precisely (White and Steig 1998). New Zealand and Patagonia both contain relatively small mountain glaciers that advanced and retreated in the late Pleistocene in response to cooling and warming climates. Evidently, these glaciers did not advance at the start of the Younger Dryas (Barrows et al. 2007; Sugden et al. 2005; Turner et al. 2005). Instead, their growth began about 300 to 500 years earlier, during a Southern Hemisphere cooling episode known as the Antarctic Cold Reversal (ACR). In Antarctica, there is no equivalent to the abrupt Bølling warming of the north. Instead, warming after the last glacial maximum (LGM) began around 18,000 years ago, and continued in a gradual, ramp-like progression thereafter, with only minor cold intervals. The ACR (which seems to have been much less severe than the Younger Dryas) appears to have been contemporaneous with the warm Allerød of the north; it began about 14,000 cal BP and ended soon after the YD began (Morgan et al. 2002). Some records from Patagonia do indicate a brief dry interval, about 12,800–12,600 cal BP, that is synchronous with the YD onset (Boës and Fagel 2008).

Generally during the late Pleistocene, warm periods (interstadials) in Antarctica were synchronous with cold periods (stadials) in the Northern Hemisphere, and vice versa (Morgan et al. 2002). This anti-phasing of climate in north and south, the so-called see-saw effect, fits well with the thermohaline circulation model: when the deep ocean currents carried warmth northward, they left a colder ocean in the southern Atlantic, which affected the adjacent landmasses.

With respect to the mystery of megafaunal extinction, the Southern Hemisphere record offers several instructive contrasts to the northern sequence. Warming, culminating in full Holocene conditions, began much earlier in the south; but there was no die-off of large mammals in southern South America at 18,000 cal BP. They did not disappear at 13,800 cal BP, when the ACR began. Instead, they survived until about 12,500 cal BP—several hundred years after

the arrival of human hunters, and just before the moderate warming that resumed at about 12,200 cal BP.

BLACK MATS

The impact theorists have been closely examining samples from "black mats" in the Southwest (Firestone et al. 2007a). Vance Haynes (2008) recently reviewed the characteristics and distribution of these features. A most important example occurs at Murray Springs, a Clovis campsite in the San Pedro Valley of southern Arizona (Haynes 2007). Here, the "mat" covers a mammoth butchered by Clovis hunters. Multiple radiocarbon dates for the Clovis occupation surface average about 10,900 BP. Given the lack of weathering of the bones, the mammoth's carcass cannot have been exposed for longer than a few years before the mat buried it. The structural and chemical signatures of this and similar mats remain ambiguous; Haynes' best guess is that they are primarily composed of dried algae. It is important to observe that these YD-onset mats, though particularly prominent, are not unique to this period; both earlier and later mats are known (Quade et al. 1998). If the mats are of algal origin, we must infer that the YD onset in the Southwest was marked by a sudden increase in rainfall and possibly decreased evaporation due to lowered air temperature. The water table rose; ponds filled former stream beds; algal pond scum floated on their surfaces; and when the ponds dried up, the algae settled on the bottom. The inferred sudden shift in rainfall patterns must be attributed to the changed behavior of the air masses carrying moisture from the Pacific coast. This raises again the question of the mechanism that rapidly propagated an atmospheric or oceanic signal of the YD onset from the North Atlantic region to the Pacific (Mikolajewicz et al. 1997). Sediment cores from the Santa Barbara Basin show temperature changes in the ocean that appear virtually synchronous with North Atlantic events (Hendy et al. 2002).

SANTA ROSA ISLAND

The Santa Barbara area offers a crucial test case for rival extinction theories. A dwarfed species of mammoth (*Mammuthus exilis*) had flourished on the Channel Islands off southern California (then conjoined as a single mass called Santarosae) for more than 35,000 years, but the last known specimen dates to 11,030±50 BP (Agenbroad et al. 2005). This animal's death was synchronous both with the arrival of humans and a conflagration that permanently destroyed the island's coniferous forests. The indisputable evidence of human presence is the burial of Arlington Springs Man, consisting only of a few leg bones. Originally discovered by Phil Orr in 1960, the burial site was relocated and excavated several years ago (Johnson et al. 2008). Dates have been obtained directly on the human bones (10,960±80 BP), on nearby mouse bones, and on charcoal just above the burial (11,250±40 BP). A black layer was depos-

ited just above the bones. Although its stratigraphic position may suggest a close resemblance to Haynes' black mats, this layer actually appears to consist largely of charcoal, not algal residues. A radiocarbon date for the base of this layer is 10,725±40 BP; the top dates to 10,315±40 BP.

No impact-derived microparticles were observed in the black layer at this site; however, 1.2 km upstream, in Arlington Canyon at the AC-003 location, carbon spherules were recovered from a layer of dark blue-grey mud/silt at the base of a 5-meter thick colluvial deposit (Kennett et al. 2008a). Ten ^{14}C dates were obtained for this basal layer, ranging from 10,860±70 to 11,440±90; most ages cluster around 11,100–11,200 BP. However, along with dates of 11,235±25 and 11,375±25 BP on wood, Kennett et al. (2008a) regard ages of 11,110±35 BP on a carbon breccia "elongate" particle, 11,185±30 BP on glassy carbon, and the 11,440 BP date (on a carbon spherule!) as excessive owing to an "old wood" effect. Thus, they attempt to discount evidence that would appear to seriously undermine the impact hypothesis. If carbon spherules are of extraterrestrial origin (some, from AC-003 and other locations, are reported to contain nano-diamonds [Kennett et al. 2008b]), they cannot consist of old wood. Unless these carbon particles are derived from the pith of very old trees (which admittedly is possible), they would appear to date the massive wildfire on Santa Rosa to about 11,100–11,400 BP, or ca. 13,000–13,400 cal BP.

If the burn was sparked by a comet impact, that event must have occurred at least 100—perhaps even 500—years before the YD onset. YD-associated dates in Europe rarely exceed 10,950–11,000 BP. It is remotely possible that there is some sort of geographic offset between Pacific Coast and European tree dates (comparable to the ca. 20-year disparity between German and Anatolian trees ca. 1500 cal BC [Kromer et al. 2001]). However, barring such complications, the obvious inference is that the wildfire on Santa Rosa raged long before the start of the Younger Dryas.

Charcoal and wood samples taken from upper levels of the 5-meter deposit at AC-003 yielded dates essentially the same as those from the lowest layer—about 11,100 BP. Kennett et al.'s (2008a) explanation of this sequence is basically the same as that of Pinter and Anderson (2006). The wildfire destroyed most of the pine-juniper forest on the island. The fire was followed by heavy rains that carved out the canyon; then, periodically, chunks of eroded sediment, carrying embedded charred remnants of the trees, slumped into the channel.

However, Pinter (pers. comm. 2008) does not accept an extraterrestrial cause for the wildfire. He suspects, and I agree, that humans probably set the fire. Kennett et al. (2008a) deploy the usual arguments against human involvement that have been invoked regularly against Paul Martin's overkill theory (e.g., Grayson and Meltzer 2002, 2003; *contra* Fiedel and Haynes 2004): no kill/butchery sites or Clovis points have been found on Santa Rosa, and not many fluted points are known from the adjacent southern California mainland. They also observe that a gap of seven centuries appears to intervene between Arlington Springs Man and the next known human presence on the islands. Their expectation of kill sites is unrealistic. Some 40,000 years of mammoth occupation of the island are represented by only 140 fossil

localities; in other words, one carcass survived as a fossil every 300 years. The presumed kill-off at ca. 13,000 cal BP may have been a matter of just a few years or even weeks. Given the now-documented heavy erosion of the landscape following the great fire, the dismembered carcasses on occupied surfaces would have been swept away and dispersed. While it is true that no Clovis points have been found on Santa Rosa, five tiny biface-reduction flakes of nonlocal chert were excavated at the site of the human leg bones. The associates of Arlington Springs Man evidently were armed with some sort of bifacial projectile points; and Clovis-like points actually form a small cluster on the nearby mainland (Kennett et al. 2008a: Figure 1A). Of course, the human skeletal remains alone offer unequivocal proof of human presence around 13,200 cal BP; and most crucially, the dates for the wildfire show that it occurred a century or two before the Younger Dryas. It therefore cannot be credibly ascribed either to sudden climate change or to the hypothesized 12,900 cal BP impact.

The particular case of the wildfire on Santa Rosa may be telling us something important about the broader, continental-scale record of initial human colonization. In Australia, the first humans, arriving around 46,000 cal BP (Gillespie 2008), apparently started major fires, either accidentally or intentionally. These fires dramatically changed the vegetation cover, stressed the native fauna (leading to extinction of many species), and through atmospheric feedback mechanisms, may even have permanently changed rainfall patterns (Miller et al. 1999, 2005a, 2005b). The arrival of Polynesian ancestors on New Zealand and other Pacific islands is regularly marked by evidence of burning (Holdaway and Jacomb 2000; Kirch 1996); similarly, the first settlers of Madagascar set fires (Burney et al. 2003). In all these cases, faunal extinctions also quickly followed human arrival.

In the Americas, with dramatic evidence of direct human predation on megafauna, we have rarely considered the environmental havoc that Paleoindians may have wreaked as fire-starters. However, there are hints of a pattern. In northern Alaska (where still-undated small fluted points indicate a Paleoindian presence), fires start to occur—with no evidence of aridity that might have encouraged natural burns—about 14,000 cal BP (Higuera et al. 2008)—coeval with the earliest evidence of human settlement in central Alaska (Crass and Holmes 2004). Horses went extinct in Alaska very soon after humans arrived, followed by mammoths a few centuries later (Guthrie 2006; Solow et al. 2006; Buck and Bard 2007). In southeastern New York, a major reduction in the population of Late Glacial mastodon and other megafauna is indicated by a fall-off in the spores of *Sporormiella* fungus in pond sediments (Robinson et al. 2005); this fungus thrived on megamammal dung (Davis 1987; Davis and Shafer 2006).

Following the fungus decline, a rise in charcoal particles attests to local fires. Direct dates then document total extinction of mastodon by about 12,800 cal BP. Robinson et al. (2005) suggest that this sequence is attributable to human hunting, although they assume the fires, fueled by vegetation accumulating due to depletion of the megaherbivore population, were naturally

ignited. The arrival of Paleoindians in Central America seems to be marked by charcoal particles in Panama at ca. 11,000 BP (Piperno et al. 1990). In Patagonia, fluted Fell I fishtail points mark Paleoindian arrival at 11,000 BP (Jackson et al. 2007; Steele and Politis 2008). Charcoal shows wildfires beginning in the region at 12,800 cal BP (Massaferro et al. 2009)—coincident with human arrival, and not with the preceding climatic stresses of the Antarctic Cold Reversal or Huelmo/Mascardi cold interval (which began ca. 13,300 cal BP [Hajdas et al. 2003; Boës and Fagel 2008]).

The Late Glacial fire record across North America, as tracked by charcoal particles in lake sediments, has been summarized recently by a large team of paleoecologists who were explicitly testing the impact hypothesis (Marlon et al. 2009). They find no dramatic peak in fire incidence at 12,900 cal BP. Generally, they attribute increased fire frequency to post-glacial warming, but there are peak incidences at about 13,900 cal BP (presumably associated with the Older Dryas cold event), 13,200 cal BP, and 11,700 cal BP (end of the Younger Dryas). Marlon et al. (2009) admit the possibility that the fire spike evident at 13,200 cal BP might be attributable to human activity—this date would correlate very well with the initial expansion of the Clovis complex. Their alternative attribution to a rather minor cold event at that time, the Intra-Allerød Cold Period, is not very convincing.

MEGAFAUNAL EXTINCTION—SUDDEN BUT STAGGERED

Despite my skepticism about the comet impact hypothesis, I completely agree with its advocates on one crucial fact: the die-off of North American megafauna was dramatic, abrupt, and basically synchronous. At least 17 genera, and probably 34, went extinct within a mere four centuries (13,200–12,800 cal BP) (Fiedel and Haynes 2004; Fiedel 2009). Nothing like this had happened during previous glacial-interglacial cycles. This extinction happens to be exactly coeval with the florescence of the Clovis culture, which probably manifests the spread of the first humans across the continent. This is unlikely to be a mere coincidence. I suspect that hunting and environmental transformation by humans—possibly including the use of fire—were primarily responsible for the extinctions. In this respect, both the overkill and impact theories stand in sharp contrast to the climate-induced extinction model that postulates a gradual extinction process extended across millennia (Grayson 2007; Grayson and Meltzer 2002, 2003).

Admittedly, the incongruous survival of diverse species confounds both the overkill and impact scenarios. Why didn't the thermal shock wave that fried mammoths, mastodons, short-faced bears, ground sloths, giant beavers, and flat-headed peccaries also incinerate the indigenous deer, bison, black and brown bears, as well as the just-arrived elk and moose? One might as easily inquire why bison, demonstrably often hunted by Paleoindians, survived the supposed overkill, while horses did not. If Paleoindians were responsible for the abrupt demise of the pygmy mammoths of Santa Rosa, why did it

take so long for continual human predation to wipe out the island's flightless ducks, which survived into the mid-Holocene (Jones et al. 2008)?

Overkill has some untidy loose ends, but so does the impact theory. If the heat wave generated by a blast in eastern Canada was powerful enough to ignite fires on Santa Rosa Island and immolate the dwarf *Mammuthus exilis* there, why did it not also finish off the small mammoths of Wrangel Island off Siberia, or those living on the Pribilof Islands off Alaska, which evidently survived into the mid-Holocene (Vartanyan et al. 1995; Arslanov et al. 1998; Guthrie 2004; Crossen et al. 2005)? The medium-sized ground sloths of the Caribbean islands also survived into the mid- or late Holocene, even though the megafauna of Florida, only 100 miles away, died off along with other mainland animals at the YD onset (Steadman et al. 2005; MacPhee et al. 2007). In this case, the explanation seems easy: humans did not get to the islands until about 6000 cal BP, and the larger mammals began to die off only then—an example of Paul Martin's "deadly syncopation", also evident in Australia, New Zealand, and Madagascar (Martin 1967, 2005; Martin and Stuart 1995; Holdaway and Jacomb 2000; Burney et al. 2003; Miller et al. 1999, 2005a; Gillespie 2008).

Farther south, the extinction of South American megafauna (some 50 genera, compared to the 30 that disappeared from the north [Cione et al. 2003]), appears to have been slightly later than that of the northern animals. Radiocarbon dates of ca. 10,400–10,200 BP (e.g., Martinez 2001; Long et al. 1998) suggest that many species survived until about 12,000 cal BP, and some may even have lasted for centuries after that (Garcia 2003; Suarez 2003; Politis et al. 2004; Fiedel 2009). It seems unlikely that a bolide impact could have produced such a staggered effect. Of course, the delayed extinction also creates some difficulty for the pure overkill scenario, as it seems that Clovis descendants arrived at Tierra del Fuego no more than a century or so after Clovis expansion in North America (Fiedel 2006; Waters and Stafford 2007). I suspect that the differing impacts of human hunting might have something to do with differences in the indigenous predator guilds; apart from puma and jaguar, which survived into the Holocene, the only big feline predator in Late Glacial South America was *Smilodon populator* (Neves and Pilo 2003). Before humans arrived, megaherbivore populations may have been limited only by the availability of edible plants, not by heavy predation as in North America. Importantly, the onset of Holocene climate was both earlier and less abrupt in South America, and the Younger Dryas–like Antarctic Cold Reversal preceded the YD; thus, climate change does not appear responsible for the later extinctions in the southern continent.

In contrast to the Caribbean and South America, extinctions in Beringia evidently *preceded* the supposed comet impact: horses seem to have died out around 14,000 cal BP, and mammoth around 13,300 cal BP (Solow et al. 2006; Buck and Bard 2007). Although environmental changes set in motion by Bølling warming may have been implicated (Guthrie 2006), the earliest human campsite in Alaska, Swan Point, dates to about 14,300 cal BP (Crass and Holmes 2004), just prior to the disappearance of horses—and horse bones have been found at that site (C. Holmes pers. comm. 2007).

No Discontinuity of Human Occupation in the Americas

The impact advocates' assertion that the Paleoindian bearers of the Clovis culture were nearly eradicated by the hypothesized event (e.g., Anderson et al. 2008; Culleton and Kennett 2008) is an egregious error (Fiedel 2007; Buchanan et al. 2008). Clovis did not disappear at the YD onset; rather, it was transformed into a series of more regionally specific cultures that continued to make fluted points of distinct styles: Folsom in the Plains, Dalton in the Midwest, Debert in the Northeast. Numerous surveys in the Plains and Southwest have shown that Folsom points and sites far outnumber Clovis finds (e.g., Judge 1973). Even if we attribute a longer temporal span to Folsom (say 12,800 to 11,800 cal BP), that does not account for so great a disparity. Dalton points similarly outnumber Clovis in the Midwest, and in Florida, Suwannee and Simpson points far outnumber earlier Clovis. In the Northeast, classic Clovis points are rare or absent. However, points with exaggerated basal concavities, obviously derived from Clovis, were dated to about 10,700–10,400 BP (i.e., early Younger Dryas age) at Debert in Nova Scotia (MacDonald 1968). Similar points of the same age have been found at Vail and other sites in Maine, and another derivative fluted form, Barnes points, were made by Paleoindians in the Great Lakes region during the Younger Dryas. If the supposed impact was centered on eastern Canada, one must ask, why did Paleoindians of the Clovis tradition immediately trek into the region surrounding the impact site and what surviving fauna did they find there to eat?

Perhaps the Clovis extinction scenario tied to the comet has some appeal to those who would seek European ancestry for so-called Paleoamericans (e.g., Stanford and Bradley 2002; Bradley and Stanford 2004), as this would allow a complete population replacement by some archaeologically invisible Asian-descended immigrants from Beringia. How else can they explain the unambiguously Central Siberian character of modern Native Americans' genes (e.g., Zegura et al. 2004)? But, in fact, Paleoindian cultures made a gradual transition into the various regional Early Archaic cultures (with some undoubted redistribution of people on the landscape as environments changed); there was no population replacement and so no need to invoke an extraterrestrial *deus ex machina*.

The Geologists Demur

Geologists who specialize in impact phenomena seem generally unimpressed with the nanodiamonds, carbon spherules, etc. (Pinter and Ishman 2008; Kerr 2007, 2009). They observe that the Earth is constantly bombarded by a rain of micrometeorites. Vance Haynes has successfully collected microparticles from the roof of his house (Haynes pers. comm. 2008). West and Kennett (2008) counter that the quantities of spherules and rare elements derived from the "YDB" (Younger Dryas Boundary) zone greatly exceed, by

orders of magnitude, the usual background levels. Skeptics respond: how can the impact team be sure that the alleged concentration does not result from long exposure of the surface? Given the supposed extension of the impact's effects into Europe (Firestone et al. 2007a, but contested by van der Hammen and van Geel 2008), it should be fairly easy to resolve this particular issue by examination of annually varved sediments in central European lakes. Extraterrestrial microparticles should be highly concentrated in the varves dating to the few years of YD onset, and absent or very rare above and below them. Two recent studies have failed to replicate the claimed concentration of microspherules and rare elements (iridium and osmium) in YDB deposits (Surovell et al. 2009; Paquay et al. 2009). West and Kennett now claim (in a March 2009 episode of the PBS TV science program, *NOVA*) that extraordinary quantities of extraterrestrial nanodiamonds are present in a new sample of Greenland ice dated to the YD onset; a more formal report in *Proceedings of the National Academy of Sciences* is said to be forthcoming (E. Hadingham, pers. comm. 2009). On the other hand, the iridium level from the same ice sample is reported to be only about three times the usual background level—not a significant spike. However, even if microparticles are shown to be as concentrated as claimed, an alternative explanation is available: cyclical peaks in cosmic dust accumulation.

Franzen and Cropp (2007) have reported a cyclical recurrence of relatively high frequencies of iron spherules and rare elements throughout the Holocene, preserved within raised peat bogs in Sweden, Ireland, and Tierra del Fuego (Argentina). The peak frequencies occur roughly every 1,250 years and correspond closely to the timing of cold Bond events recorded in North Atlantic sediments as maxima in ice-rafted debris. Both cycles appear to be related in some way, not yet fully explicated, to fluctuating solar radiation and perhaps also to variable cosmic dust flux, responsive to solar activity. The oldest bog sampled by Franzen and Cropp (2007) is Harberton Bog in Tierra del Fuego, with deposits extending as far back as 14,000 cal BP. It is particularly noteworthy that they report no spherule or rare-earth element peak there at 12,900 cal BP, corresponding to the YD onset in the Northern Hemisphere. Also notable is a generally high frequency of rare elements in the latest Pleistocene and early Holocene, peaking at ca. 8500 cal BP and then dropping by about 50 percent to near-modern amounts at 8000 cal BP. Franzen and Cropp (2007) suggest that some of the climatic effects of cosmic dust fluxes may be modulated by abrupt phytoplankton blooms, as marine algae respond rapidly (a matter of weeks) to newly available iron particles in the cosmic dust.

As I am skeptical of the impact theory on the grounds of ostensibly contradictory archaeological evidence (*contra* Culleton and Kennett 2008; Anderson et al. 2008), I obviously endorse the skeptical reactions of most geologists. Nevertheless, we may all be completely wrong. It took fifty years for Wegener's continental drift theory to win credence. The Alvarez's asteroid impact theory of dinosaur extinction was met with similar incredulity when first proposed in 1980. However, it is widely accepted today, although some geologists advocate massive volcanic activity in India as a primary or

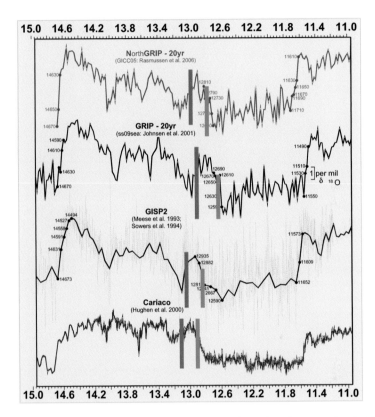

Figure 1. Younger Dryas onset and inferred date of Laacher See eruption in Greenland ice cores and Cariaco Basin varves. Green bar marks onset of YD; red bar
marks the Laacher See eruption (190 years pre-YD). After online PowerPoint presentation on Late glacial chronology by O. Jöris and B. Weninger, with permission.

complementary cause of extinction (e.g., Keller et al. 2008), and others report problems with the relative timing of the impact and the extinctions (e.g., Keller 2003). I have not yet heard a convincing alternative explanation from geologist colleagues of the remarkably high concentrations of spherules, helium 3, and iridium that are reported in 12,900-year-old YD onset layers, including the "black mats" of the Southwest. However, I note that two substantial volcanic eruptions occurred about two hundred years before the YD onset—Glacier Peak in Oregon (Mehringer et al. 1984) and Laacher See in Germany (Litt et al. 2003) (Fig. 1)—and I wonder if these might account for the iridium anomaly.

Finally, in assessing the impact theory, we must beware of the logical fallacy of *post hoc ergo propter hoc*. The Tunguska impact of 1908 was followed only a decade later by the Russian Revolution. Despite the near-contemporaneity of these events, as far as I am aware no one has suggested that the well-documented destruction of thousands of Siberian trees created an ecological catastrophe that motivated Russian workers and peasants to overthrow the Romanoffs. My point is that the various microparticles may indeed provide convincing evidence of an impact at 12,900 cal BP, but the climate changes, megafauna extinctions, and human cultural transformations that occurred around the same time were not necessarily caused by that impact. Jim Kennett and I agreed in a recent conversation that the South American geological and faunal records will provide a crucial test of the impact hypothesis, and I hope that we will see that test undertaken in the near future.

REFERENCES

Agenbroad, L. D., J. Johnson, D. Morris, and T. W. Stafford Jr. 2005. Mammoths and humans as Late Pleistocene contemporaries on Santa Rosa Island. In *Proceedings of the sixth California Islands symposium, Ventura, California, December 1–3, 2003*, ed. D. K. Garcelon and C. A. Schwemm, 3–7. Arcata, Calif.: Institute for Wildlife Studies.

Alley, R. B., D. A. Meese, C. A. Shuman, A. J. Gow, K. C. Taylor, P. M. Grootes, J. W. C. White, M. Ram, E. D. Waddington, P. A. Mayewski, and G. A. Zielinski. 1993. Abrupt increase in Greenland snow accumulation at the end of the Younger Dryas event. *Nature* 362: 527–29.

Anderson, D. G., S. C. Meeks, D. S. Miller, S. J. Yerka, J. C. Gillam, A. C. Goodyear, E. N. Johanson, and A. West. 2008. The effect of the Younger Dryas on Paleoindian occupations in eastern North America: evidence from artifactual, pollen, and radiocarbon records. American Geophysical Union, fall meeting 2008, abstract #PP13C-1475.

Arslanov, Kh. A., G. T. Cook, S. Gulliksen, D. D. Harkness, T. Kankainen, E. M. Scott, S. Vartanyan, and G. I. Zaitseva. 1998. Consensus dating of mammoth remains from Wrangel Island. *Radiocarbon* 40 (1–2): 289–94.

Barrows, T. T., S. J. Lehman, L. K. Fifield, and P. De Deckker. 2007. Absence of cooling in New Zealand and the adjacent ocean during the Younger Dryas chronozone. *Science* 318: 86–89.

Björck, S., B. Kromer, S. Johnsen, O. Bennike, D. Hammarlund, G. Lemdahl, G. Possnert, T. L. Rasmussen, B. Wohlfarht, C. U. Hammer, and M. Spurk. 1996. Synchronized terrestrial-atmospheric deglacial records around the North Atlantic. *Science* 274: 1150–60.

Boës, X., and N. Fagel. 2008. Timing of the Late Glacial and Younger Dryas cold reversal in southern Chile varved sediments. *Journal of Paleolimnology* 39 (2): 267–81.

Bond, G., B. Kromer, J. Beer, R. Muscheler, M. N. Evans, W. Showers, S. Hoffmann, R. Lotti-Bond, I. Hajdas, and G. Bonani. 2001. Persistent solar influence on North Atlantic climate during the Holocene. *Science* 294: 2130–36.

Bradley, B., and D. Stanford. 2004. The North Atlantic ice-edge corridor: a possible Paleolithic route to the New World. *World Archaeology* 36: 459–78.

Brauer, A., G. H. Haug, P. Dulski, D. M. Sigman, and J. F. W. Negendank. 2008. An abrupt wind shift in western Europe at the onset of the Younger Dryas cold period. *Nature Geoscience* 1: 520–23.

Broecker, W. S. 1991. The great ocean conveyor. *Oceanography* 4: 79–89.

———. 1997. Thermohaline circulation, the Achilles Heel of our climate system: will man-made CO_2 upset the current balance? *Science* 278: 1582–88.

Broecker, W. S., D. M. Peteet, and D. Rind. 1985. Does the ocean-atmosphere system have more than one stable mode of operation? *Nature* 315: 21–25.

Broecker, W. S., J. P. Kennett, B. P. Flower, J. T. Teller, S. Trumbore, G. Bonani, and W. Wolfli. 1989. Routing of meltwater from the Laurentide ice sheet during the Younger Dryas cold episode. *Nature* 341: 318–21.

Buchanan, B., M. Collard, and K. Edinborough. 2008. Paleoindian demography and the extraterrestrial impact hypothesis. *Proceedings of the National Academy of Sciences of the United States of America* 105 (33): 11651–54.

Buck, C. E., and E. Bard. 2007. A calendar chronology for Pleistocene mammoth and horse extinction in North America based on Bayesian radiocarbon calibration. *Quaternary Science Reviews* 26 (17–18): 2031–35.

Burney, D. A., G. S. Robinson, and L. P. Burney. 2003. *Sporormiella* and the late Holocene extinctions in Madagascar. *Proceedings of the National Academy of Sciences* 100 (19): 10800–805.

Carlson, A. E., P. U. Clark, B. A. Haley, G. P. Klinkhammer, K. Simmon, E. J. Brook, and K. J. Meissner. 2007. Geochemical proxies of North American freshwater routing during the Younger Dryas cold event. *Proceedings of the National Academy of Sciences of the United States of America* 104: 6556–61.

Cione, A. L., E. P. Tonni, and L. Soibelzon. 2003. The broken zig-zag: late Cenozoic large mammal and tortoise extinction in South America. *Revista Museo Argentino de Ciencia Natural* 5 (1): 1–19.

Clement, A. C., M. A. Cane, and R. Seager. 2001. An orbitally driven tropical source for abrupt climate change. *Journal of Climate* 14 (11): 2369–75.

Crass, B. A., and C. E. Holmes. 2004. Swan Point: A case for land bridge migration in the peopling of North America. Paper presented at 69th annual meeting of Society for American Archaeology, Montreal.

Crossen, K. J., D. R. Yesner, D. W. Veltre, and R. W. Graham. 2005. 5,700-year-old mammoth remains from the Pribilof Islands, Alaska: last outpost of

North American megafauna. Paper presented at annual meeting of Geological Society of America, Salt Lake City (October 16–19, 2005).

Culleton, B. J., and D. J. Kennett. 2008. Evaluating the Paleoindian radiocarbon record at the onset of the Younger Dryas: sensitivity analyses and Bayesian chronology-building. American Geophysical Union, fall meeting 2008, abstract #PP23D-03.

Davis, O. K. 1987. Spores of the dung fungus *Sporormiella*: increased abundance in historic sediments and before Pleistocene megafaunal extinction. *Quaternary Research* 28: 290–94.

Davis, O. K., and D. S. Shafer. 2006. *Sporormiella* fungal spores, a palynological means of detecting herbivore density. *Paleogeography, Palaeoclimatology, Palaeoecology* 237 (1): 40–50.

de Vernal, A., C. Hillaire-Marcel, and G. Bilodeau. 1996. Reduced meltwater outflow from the Laurentide ice margin during the Younger Dryas. *Nature* 381: 774–77.

EPICA community members. 2004. Eight glacial cycles from an Antarctic ice core. *Nature* 429: 623–28.

Fairbanks, R. G., R. A. Mortlock, T.-C. Chiu, L. Cao, A. Kaplan, T. P. Guilderson, T. W. Fairbanks, and A. L. Bloom. 2005. Marine radiocarbon calibration curve spanning 10,000 to 50,000 Years B.P. based on paired ^{230}Th/^{234}U/^{238}U and ^{14}C Dates on pristine corals. *Quaternary Science Reviews* 24: 1781–96.

Fiedel, S. J. 2006. Points in time: establishing a precise hemispheric chronology for Paleoindian migrations. In *Paleoindian archaeology, a hemispheric perspective*, ed. J. E. Morrow and C. Gnecco, 21–43. Gainesville: University Press of Florida.

———. 2007. Did a bolide impact trigger the Younger Dryas and wipe out American megafauna? A skeptic's reaction to an intriguing hypothesis. Poster, American Geophysical Union, fall meeting 2007, abstract #PP43A-04.

———. 2009. Sudden deaths: the chronology of terminal Pleistocene megafaunal extinction. In *American megafaunal extinctions at the end of the Pleistocene*, ed. G. Haynes, 21–37. Netherlands: Springer.

Fiedel, S., and G. Haynes. 2004. A premature burial: comments on Grayson and Meltzer's "requiem for overkill." *Journal of Archaeological Science* 31 (1): 121–31.

Firestone, R. B., and W. Topping. 2001. Terrestrial evidence of a nuclear catastrophe in Paleoindian times. *Mammoth Trumpet* 16 (2): 9–16.

Firestone, R. B., and A. West. 2005. Evidence for the extinction of mammoths by an extraterrestrial impact event. Paper presented at Second International World of Elephants Congress, Hot Springs, South Dakota, September 22–25.

Firestone, R. B., A. West, J. P. Kennett, L. Becker, T. E. Bunch, Z. S. Revay, P. H. Schultz, T. Belgya, D. J. Kennett, J. M. Erlandson, O. J. Dickenson, A. C. Goodyear, R. S. Harris, G. A. Howard, J. B. Kloosterman, P. Lechler, P. A. Mayewski, J. Montgomery, R. Poreda, T. Darrah, S. S. Que Hee, A. R. Smith, A. Stich, W. Topping, J. H. Wittke, and W. S. Wolbach. 2007a. Evidence for an extraterrestrial impact 12,900 years ago that contributed to the

megafaunal extinctions and the Younger Dryas cooling. *Proceedings of the National Academy of Sciences* 104: 16016–21.

Firestone, R. B., A. West, Z. Revay, T. Belgya, A. Smith, and S. S. Que Hee. 2007b. Evidence for a massive extraterrestrial airburst over North America 12.9 ka ago. *Eos* (Transactions, American Geophysical Union) 88 (23), Jt. Assem. Suppl., abstract PP41A-01.

Firestone, R. B., A. West, Z. Stefanka, Z. Revay, and J. T. Hagstrum. 2007c. Micrometeorite impacts in Beringian mammoth tusks and a bison skull. American Geophysical Union, fall meeting, 2007, San Francisco.

Firestone, R. B., A. West, Z. Revay, J. T. Hagstrum, A. Smith, and S. S. Que Hee. 2008. Elemental analysis of the sediment, magnetic grains and microspherules from the Younger Dryas impact layer. American Geophysical Union, fall meeting 2008, abstract #PP13C-1472.

Franzen, L. G., and R. A. Cropp. 2007. The peatland/ice age hypothesis revised, adding a possible glacial pulse trigger. *Geografiska Annaler: Series A, Physical Geography* 89 (4): 301–30.

Garcia, A. 2003. On the coexistence of man and extinct Pleistocene megafauna at Gruta del Indio (Argentina) 9000 C-14 years ago. *Radiocarbon* 45: 33–39.

Gillespie, Richard. 2008. Updating Martin's global extinction model. *Quaternary Science Reviews* 27 (27–28): 2522–29.

Goslar, T., M. Arnold, E. Bard, T. Kuc, M. F. Pazdur, M. Ralska-Jasiewiczowa, K. Rozanski, N. Tisnerat, A. Walanus, B. Wicik, and K. Wieckowski. 1995. High concentration of atmospheric ^{14}C during the Younger Dryas cold episode. *Nature* 377: 414–17.

Goslar, T., M. Arnold, N. Tisnerat-Laborde, J. Czernik, and K. Więckowski. 2000a. Variations of Younger Dryas atmospheric radiocarbon explicable without ocean circulation changes. *Nature* 403: 877–80.

Goslar, T., M. Arnold, N. Tisnerat-Laborde, C. Hatté, M. Paterne, and M. Ralska-Jasiewiczowa. 2000b. Radiocarbon calibration by means of varves versus ^{14}C ages of terrestrial macrofossils from Lake Gosciaz and Lake Perespilno, Poland. *Radiocarbon* 42: 335–48.

Graham, R. W. 1998. Mammals' eye view of environmental change in the United States at the end of the Pleistocene. Paper presented at 63rd Annual Meeting of the Society for American Archaeology, Seattle.

Graham, R. W., and E. L. Lundelius Jr. 1984. Coevolutionary disequilibrium and Pleistocene extinctions. In *Quaternary extinctions: a prehistoric revolution,* ed. P. S. Martin and R. G. Klein, 223–49. Tucson: University of Arizona Press.

Graham, R. W., T. Stafford, E. Lundelius, H. Semken, and J. Southon. 2002. C-14 chronostratigraphy and litho-stratigraphy of Late Pleistocene megafauna extinctions in the New World. Paper presented at 67th Annual Meeting of Society for American Archaeology, Denver.

Graham, R. W., T. Stafford, and H. Semken. 1997. Pleistocene extinctions: chronology, non-analog communities, and environmental change. Paper presented at symposium "Humans and other catastrophes," American Museum of Natural History, April.

Grayson, D. K. 2007. Deciphering North American Pleistocene extinctions. *Journal of Anthropological Research* 63: 185–213.

Grayson, D. K., and D. J. Meltzer. 2002. Clovis hunting and large mammal extinction: a critical review of the evidence. *Journal of World Prehistory* 15: 1–68.

———. 2003. A requiem for North American overkill. *Journal of Archaeological Science* 30: 585–93.

Guthrie, R. D. 2004. Radiocarbon evidence of mid-Holocene mammoths stranded on an Alaskan Bering Sea island. *Nature* 429: 746–49.

———. 2006. New carbon dates link climatic change with human colonization and Pleistocene extinctions. *Nature* 441: 207–9.

Hajdas, I., G. Bonani, P. Bodén, D. M. Peteet, and D. H. Mann. 1998. Cold reversal on Kodiak Island, Alaska, correlated with the European Younger Dryas by using variations of atmospheric ^{14}C content. *Geology* 26: 1047–50.

Hajdas, I., G. Bonani, P. I. Moreno, and D. Ariztegui. 2003. Precise radiocarbon dating of Late-Glacial cooling in mid-latitude South America. *Quaternary Research* 59: 70–78.

Haynes, C. V., Jr. 2007. Nature and origin of the black mat, stratum F2. In *Murray Springs*, ed. C. V. Haynes Jr. and B. Huckell, Appendix B, 240–49. Tucson: University of Arizona Press.

———. 2008. Younger Dryas "black mats" and the Rancholabrean termination in North America. *Proceedings of the National Academy of Sciences* 105: 6520–25.

Hendy, I. L., J. P. Kennett, E. B. Roark, and B. L. Ingram. 2002. Apparent synchroneity of submillennial scale climate events between Greenland and Santa Barbara Basin, California from 30–10 ka. *Quaternary Science Reviews* 21: 1167–84.

Higuera, P. E., L. B. Brubaker, P. M. Anderson, T. A. Brown, A. T. Kennedy, and F. S. Hu. 2008. Frequent fires in ancient shrub tundra: Implications of paleo-records for Arctic environmental change. *PLoS ONE* 3: e0001744.

Hoffman, Carey. 2008. Exploding asteroid theory strengthened by new evidence located in Ohio, Indiana. *University of Cincinnati News*, July 2, 2008, online at http://www.uc.edu/news/NR.asp?id=8625 [last accessed 01/17/10].

Holdaway, R. N., and C. Jacomb. 2000. Rapid extinction of the moas (Aves: Dinornithiformes): model, test, and implications. *Science* 287: 2250–54.

Hua, Q., M. Barbetti, D. Fink, K. F. Kaiser, M. Friedrich, B. Kromer, V. A. Levchenko, U. Zoppi, A. M. Smith, and F. Bertuch. 2009. Atmospheric ^{14}C variations derived from tree rings during the early Younger Dryas. *Quaternary Science Reviews*, doi:10.1016/j.quascirev.2009.08.013 (in press).

Hughen, K. A., J. R. Southon, S. J. Lehman, and J. T. Overpeck. 2000. Synchronous radiocarbon and climate shifts during the last deglaciation. *Science* 290: 1951–54.

Hughen, K. A., J. R. Southon, C. J. H. Bertrand, B. Frantz, and P. Zermeño. 2004. Cariaco Basin calibration update: revisions to calendar and ^{14}C chronologies for core PL07-58 pc. *Radiocarbon* 46: 1161–87.

Ivester, A. H., D. I. Godfrey-Smith, M. J. Brooks, and B. E. Taylor. 2004. The timing of Carolina Bay and inland activity on the Atlantic coastal plain of Georgia and South Carolina. *Geological Society of America Abstracts with Programs* 36 (5): 69.

Jackson, D., C. Mendez, R. Seguel, A. Maldonado, and G. Vargas. 2007. Initial occupation of the Pacific Coast of Chile during Late Pleistocene times. *Current Anthropology* 48 (5): 725–31.

Johnsen, S. J., H. B. Clausen, W. Dansgaard, K. Fuhrer, N. Gundestrup, C. U. Hammer, P. Iversen, J. Jouzel, B. Stauffer, and J. P. Steffensen. 1992. Irregular glacial interstadials recorded in a new Greenland ice core. *Nature* 359: 311–13.

Johnsen, S. J., D. Dahl-Jensen, N. Gundestrup, J. P. Steffensen, H. B. Clausen, H. Miller, V. Masson-Delmotte, A. E. Sveinbjörndottir, and J. White. 2001. Oxygen isotope and palaeotemperature records from six Greenland ice-core stations: Camp Century, Dye-3, GRIP, GISP2, Renland, and North-GRIP. *Journal of Quaternary Science* 16: 299–307.

Johnson, J. R., T. W. Stafford Jr., G. J. West, and T. K. Rockwell. 2008. Environmental change at Arlington Springs before and after the Younger Dryas. Paper presented at the Society for American Archaeology Annual Meeting, Vancouver, March 28, 2008.

Jones, T. L., J. F. Porcasi, J. M. Erlandson, H. Dallas Jr., T. A. Wake, and R. Schwaderer. 2008. The protracted Holocene extinction of California's flightless sea duck (*Chendytes lawi*) and its implications for the Pleistocene overkill hypothesis. *Proceedings of the National Academy of Sciences* 105: 4105–8.

Judge, W. J. 1973. *Paleoindian occupation of the central Rio Grande Valley in New Mexico*. Albuquerque: University of New Mexico Press.

Keller, G. 2003. Biotic effects of impacts and vulcanism. *Earth and Planetary Science Letters* 215: 249–64.

Keller, G., T. Adatte, S. Gardin, A. Bartolini, and S. Bajpai. 2008. Main Deccan phase of volcanism ends near K-T boundary: evidence from the Krishna-Godavari Basin, SE India. *Earth and Planetary Science Letters* 268: 293.

Kennett, D. J., J. P. Kennett, G. J. West, J. M. Erlandson, J. R. Johnson, I. L. Hendy, A. West, B. J. Culleton, T. L. Jones, and T. W. Stafford Jr. 2008a. Wildfire and abrupt ecosystem disruption on California's Northern Channel Islands at the Ållerød–Younger Dryas boundary (13.0–12.9 ka). *Quaternary Science Reviews* 27 (27–28): 2530–45.

Kennett, D. J., J. P. Kennett, A. West, G. J. West, T. E. Bunch, B. J. Culleton, J. M. Erlandson, S. S. Que Hee, J. R. Johnson, C. Mercer, M. Sellers, T. W. Stafford, A. Stich, J. C. Weaver, J. H. Wittke, and W. S. Wolbach. 2008b. Impact-shocked diamonds, abrupt ecosystem disruption, and mammoth extinction on California's Northern Channel Islands at the Allerod-Younger Dryas boundary (13.0–12.9 ka). American Geophysical Union, fall meeting 2008, abstract #PP23D-04D.

Kennett, D. J., J. P. Kennett, A. West, C. Mercer, S. S. Que Hee, L. Bement, T. E. Bunch, M. Sellers, and W. S. Wolbach. 2009. Nanodiamonds in the Younger Dryas boundary sediment layer. *Science* 323: 94.

Kerr, R. A. 2007. Mammoth-killer impact gets mixed reception from earth scientists. *Science* 316: 1264–65.

———. 2009 . Did the mammoth slayer leave a diamond calling card? *Science* 323: 26.

Kirch, P. V. 1996. Late Holocene human-induced modifications to a central Polynesian island ecosystem. *Proceedings of the National Academy of Sciences* 93: 5296–5300.

Kobres, R., G. A. Howard, A. West, R. B. Firestone, J. P. Kennett, D. Kimbel, and W. Newell. 2007. Formation of the Carolina Bays: ET impact vs. wind-and-water. Poster paper PP43A-10, May 2007 AGU Conference, Acapulco, Mexico.

Kromer, B., S. W. Manning, P. I. Kuniholm, M. W. Newton, M. Spurk, and I. Levin. 2001. Regional $^{14}CO_2$ offsets in the troposphere: magnitude, mechanisms, and consequences. *Science* 294: 2529–32.

Kromer, B., M. Friedrich, K. A. Hughen, F. Kaiser, S. Remmele, M. Schaub, and S. Talamo. 2004. Late glacial ^{14}C ages from a floating, 1382-ring pine chronology. *Radiocarbon* 46: 1203–9.

Lea, D. W., D. K. Pak, L. C. Peterson, and K. A. Hughen. 2003. Synchroneity of tropical and high-latitude Atlantic temperatures over the last glacial termination. *Science* 301: 1361–64.

Litt, T., H. U. Schmincke, and B. Kromer. 2003. Environmental response to climate and volcanic events in central Europe during the Weichselian Lateglacial. *Quaternary Science Reviews* 22: 7–32.

Long, A., P. S. Martin, and H. A. Lagiglia. 1998. Ground sloth extinction and human occupation at Gruta del Indio, Argentina. *Radiocarbon* 40: 693–700.

Lowell, T. V., N. Waterson, T. Fisher, H. Loope, K. Glover, G. Comer, I. Hajdas, G. Denton, J. Schaefer, V. Rinterknecht, W. Broecker, and J. Teller. 2005. Testing the Lake Agassiz meltwater trigger for the Younger-Dryas. *Eos* (Transactions, American Geophysical Union) 86: 365–73.

MacDonald, G. F. 1968. *Debert: a Palaeo-Indian site in central Nova Scotia*. Ottawa: National Museum of Man.

MacPhee, R. D. E., M. A. Iturralde-Vinent, and O. Jiménez Vázquez. 2007. Prehistoric sloth extinctions in Cuba: implications of a new "last" appearance date. *Caribbean Journal of Science* 43: 94–98.

Marlon, J. R., P. J. Bartlein, M. K. Walsh, S. P. Harrison, K. J. Brown, M. E. Edwards, P. E. Higuera, M. J. Power, R. S. Anderson, C. Briles, A. Brunelle, C. Carcaillet, M. Daniels, F. S. Hu, M. Lavoie, C. Long, T. Minckley, P. J. H. Richard, A. C. Scott, D. S. Shafer, W. Tinner, C. E. Umbanhowar Jr., and C. Whitlock. 2009. Wildfire responses to abrupt climate change in North America. *Proceedings of the National Academy of Sciences* 106: 2519–24.

Marchal, O., T. F. Stocker, F. Joos, A. Indermühle, T. Blunier, and J. Tschumi. 1999. Modelling the concentration of atmospheric CO_2 during the Younger Dryas climate event. *Climate Dynamics* 15:341–54.

Martin, P. S. 1967. Prehistoric overkill. In *Pleistocene extinctions: the search for a cause*, ed. P. S. Martin and H. E. Wright Jr., 75–120. New Haven: Yale University Press.

————. 2005. *Twilight of the mammoths.* Berkeley and Los Angeles: University of California Press.

Martin, P. S., and A. J. Stuart. 1995. Mammoth extinction: two continents and Wrangel Island. *Radiocarbon* 37: 7–10.

Martinez, G. A. 2001."Fish-tail" projectile points and megamammals: new evidence from Paso Otero 5 (Argentina). *Antiquity* 75: 532–38.

Massaferro, J. I., P. I. Moreno, G. H. Denton, M. Vandergoes, and A. Dieffenbacher-Krall. 2009. Chironomid and pollen evidence for climate fluctuations during the last glacial termination in NW Patagonia. *Quaternary Science Reviews* 28: 518–25.

Mehringer, P. J., Jr., J. C. Sheppard, and F. F. Foit, Jr. 1984. The age of Glacier Peak tephra in west-central Montana. *Quaternary Research* 21: 36–41.

Mikolajewicz, U., T. J. Crowley, A. Schiller, and R. Voss. 1997. Modelling teleconnections between the North Atlantic and North Pacific during the Younger Dryas. *Nature* 387: 384–87.

Miller, G. H., J. W. Magee, B. J. Johnson, M. L. Fogel, N. A. Spooner, M. T. McCulloch, and L. K. Ayliffe. 1999. Pleistocene extinction of *Genyornis newtoni:* human impact on Australian megafauna. *Science* 283: 205–8.

Miller, G. H., M. L. Fogel, J. W. Magee, M. K. Gagan, S. J. Clarke, and B. J. Johnson. 2005a. Ecosystem collapse in Pleistocene Australia and a human role in megafaunal extinction. *Science* 309: 287–90.

Miller, G. H., J. Mangan, D. Pollard, S. Thompson, B. Felzer, and J. Magee. 2005b. Sensitivity of the Australian monsoon to insolation and vegetation: implications for human impact on continental moisture balance. *Geology* 33: 65–68.

Mehringer P. J., Jr., J. C. Sheppard, and F. F. Foit Jr. 1984. The age of Glacier Peak tephra in west-central Montana. *Quaternary Research* 21: 36–41.

Morgan, V., M. Delmotte, T. van Ommen, J. Jouzel, J. Chappellaz, S. Woon, V. Masson-Delmotte, and D. Raynaud. 2002. Relative timing of deglacial climate events in Antarctica and Greenland. *Science* 297: 1862–64.

Murton, J. B., M. D. Bateman, S. R. Dallimore, J. T. Teller, and Z. Yang. 2010. Identification of Younger Dryas outburst flood path from Lake Agassiz to the Arctic Ocean. *Nature* 464 (7289): 740–43.

Muscheler, R., B. Kromer, S. Björck, A. Svensson, M. Friedrich, K. F. Kaiser, and J. Southon. 2008. Tree rings and ice cores reveal ^{14}C calibration uncertainties during the Younger Dryas. *Nature Geoscience* 1: 263–67.

Neves, W. A., and L. B. Pilo. 2003. Solving Lund's dilemma: new AMS dates confirm that humans and megafauna coexisted at Lagoa Santa. *Current Research in the Pleistocene* 20: 57–60.

Paquay, F. S., S. Goderis, G. Ravizza, F. Vanhaeck, M. Boyd, T. A. Surovell, V. T. Holliday, C. V. Haynes, Jr., and P. Claeys. 2009. Absence of geochemical evidence for an impact event at the Bølling–Allerød/Younger Dryas transition. *Proceedings of the National Academy of Sciences*, in press.

Peltier, W. R., A. de Vernal, and C. Hillaire-Marcel. 2008. Rapid climate change and Arctic Ocean freshening: comment and reply. Online at http://www.gsajournals.org/perlserv/?request=get-static&name=i0091-7613-36-10-e178 [last accessed January 17, 2010].

Pinter, N., and S. A. Anderson. 2006. A mega-fire hypothesis for latest Pleistocene paleo-environmental change on the Northern Channel Islands, California. *Geological Society of America Abstracts with Programs* 38 (7): 66–13.

Pinter, N., and S. E. Ishman. 2008. Impacts, mega-tsunami, and other extraordinary claims. *GSA Today* 18: 37–38.

Piperno, D. R., M. B. Bush, and P. A. Colinvaux. 1990. Paleoenvironments and human occupation in late-glacial Panama. *Quaternary Research* 33: 108–16.

Politis, G. G., P. G. Messineo, and C. A. Kaufmann. 2004. El poblamiento temprano de las llanuras pampeanas de Argentina y Uruguay (The early peopling of the Pampean plains of Argentina and Uruguay). *Complutum* 15: 207–24.

Quade, J., R. M. Forester, W. L. Pratt, and C. Carter. 1998. Black mats, spring-fed streams, and Late Glacial age recharge in the southern Great Basin. *Quaternary Research* 49: 129–48.

Rasmussen, S. O., K. K. Andersen, A. M. Svensson, J. P. Steffensen, B. M. Vinther, H. B. Clausen, M.-L. Siggaard-Andersen, S. J. Johnsen, L. B. Larsen, D. Dahl-Jensen, M. Bigler, R. Röthlisberger, H. Fischer, K. Goto-Azuma, M. E. Hansson, and U. Ruth. 2006. A new Greenland ice core chronology for the last glacial termination. *Journal of Geophysical Research* 111: D06102.

Renssen, H., B. van Geel, J. van der Plicht, and M. Magny. 2000. Reduced solar activity as a trigger for the start of the Younger Dryas? *Quaternary International* 68: 373–83.

Robinson, G., L. P. Burney, and D. A. Burney. 2005. Landscape paleoecology and megafaunal extinction in southeastern New York. *Ecological Monographs* 75: 295–315.

Schaefer J. M., G. H. Denton, D. J. A. Barrell, S. Ivy-Ochs, P. W. Kubik, B. G. Andersen, F. M. Phillips, T. V. Lowell, and C. Schlüchter. 2006. Near-synchronous interhemispheric termination of the last Glacial Maximum in mid-latitudes. *Science* 312: 1510–13.

Solow, A. R., D. L. Roberts, and K. M. Robbirt. 2006. On the Pleistocene extinctions of Alaskan mammoths and horses. *Proceedings of the National Academy of Science* 103: 7351–53.

Southon, J. R. 2002. A first step to reconciling the GRIP and GISP2 ice-core chronologies, 0–14500 yr B.P. *Quaternary Research* 57: 32–37.

Southon, J. R. 2007. Radiocarbon and the Younger Dryas. PowerPoint presentation online at http://ecology.botany.ufl.edu/radiocarbono8/Downloads/2007%20Lectures/2007%20Paleo2%20Southon.ppt#432, 2, Slide 2 [last accessed May 8, 2009].

Southon, J. R., and R. E. Taylor. 2002. Brief comments on "terrestrial evidence of a nuclear catastrophe in Paleoindian times." *Mammoth Trumpet* 17 (March).

Stanford, D., and B. Bradley. 2002. Ocean trails and prairie paths? Thoughts about Clovis origins. In *The first Americans: the Pleistocene colonization of the New World*, ed. N. G. Jablonski, 255–71. San Francisco: California Academy of Sciences, Memoirs No. 27.

Stanford, J. D., E. J. Rohling, S. H. Hunter, A. P. Roberts, S. O. Rasmussen, E. Bard, J. McManus, and R. G. Fairbanks. 2006. Timing of meltwater pulse

1a and climate responses to meltwater injections. *Paleoceanography* 21: PA4103, doi:10.1029/2006PA001340.

Steadman, D. W., P. S. Martin, R. D. E. MacPhee, A. J. T. Jull, H. G. McDonald, C. A. Woods, M. Iturralde-Vinent, and G. W. L. Hodgins. 2005. Asynchronous extinction of late Quaternary sloths on continents and islands. *Proceedings of the National Academy of Sciences* 102: 11763–68.

Steele, J., and G. Politis. 2008. AMS ¹⁴C dating of early human occupation of southern South America. *Journal of Archaeological Science* 36: 419–29.

Steffensen, J. P., K. K. Andersen, M. Bigler, H. B. Clausen, D. Dahl-Jensen, H. Fischer, K. Goto-Azuma, M. Hansson, S. J. Johnsen, J. Jouzel, V. Masson-Delmotte, T. Popp, S. O. Rasmussen, R. Röthlisberger, U. Ruth, B. Stauffer, M-L. Siggaard-Andersen, Á. E. Sveinbjörnsdóttir, A. Svensson, and J. W. C. White. 2008. High-resolution Greenland ice core data show abrupt climate change happens in few years. *Science* 321: 680–84.

Suarez, R. 2003. First records of Pleistocene fauna for an archaeological context in Uruguay: evidence from Pay Paso locality, site 1. *Current Research in the Pleistocene* 20: 113–16.

Sugden, D. E., M. J. Bentley, C. J. Fogwill, N. R. J. Hulton, R. D. McCulloch, and R. S. Purves. 2005. Late-Glacial glacier events in southernmost South America: a blend of "northern" and "southern" hemispheric climate signals? *Geografiska Annaler* 82A: 273–88.

Surovell T. A., V. T. Holliday, J. A. M. Gingerich, C. Ketron, C. V. Haynes Jr., I. Hilman, D. P. Wagner, E. Johnson, and P. Claeys. 2009. An independent evaluation of the Younger Dryas extraterrestrial impact hypothesis. *Proceedings of the National Academy of Sciences*, in press.

Tarasov, L., and W. R. Peltier. 2005. Arctic freshwater forcing of the Younger Dryas cold reversal. *Nature* 435: 662–65.

———. 2006. A calibrated deglacial drainage chronology for the North American continent: evidence of an Arctic trigger for the Younger Dryas. *Quaternary Science Reviews* 25: 659–88.

Topping, W. 1999. Cosmogenic radiocarbon as a source of error at Paleo-Indian sites and evidence for a prehistoric catastrophe. September 21, 1999. http://authors.aps.org/eprint/files/1999/Sep/aps1999sep21_004/Models .htm [last accessed January 17, 2010].

Turner K. J., C. J. Fogwill, R. D. Mcculloch, and D. E. Sugden. 2005. Deglaciation of the eastern flank of the North Patagonian icefield and associated continental-scale lake diversions. *Geografiska Annaler* 87: 363–74.

van der Hammen, T., and B. van Geel. 2008. Charcoal in soils of the Allerød-Younger Dryas transition were the result of natural fires and not necessarily the effect of an extra-terrestrial impact. *Netherlands Journal of Geosciences— Geologie en Mijnbouw* 87: 359–61.

Vartanyan, S. L., Kh. A. Arslanov, T. V. Tertychnaya, and S. B. Chernov. 1995. Radiocarbon dating evidence for mammoths on Wrangel Island, Arctic Ocean, until 2000 BC. *Radiocarbon* 37: 1–6.

Wang, Y. J., H. Cheng, R. L. Edwards, Z. S. An, J. Y. Wu, C.-C. Shen, J. A. Dorale. 2001. A high-resolution absolute-dated Late Pleistocene monsoon record from Hulu Cave, China. *Science* 294: 2345–48.

Waters, M. R., and T. W. Stafford Jr. 2007. Redefining the age of Clovis: implications for the peopling of the Americas. *Science* 315: 1122–26.

Weaver, A. J., O. A. Saenko, P. U. Clark, and J. X. Mitrovica. 2003. Meltwater Pulse 1A from Antarctica as a trigger of the Bølling-Allerød warm interval. *Science* 299: 1709–13.

West, A., and J. Kennett. 2008. Special presentation on the proposed comet impact in North America at about 11,000 14C yr B.P. and its impact on people and fauna. Presented at Workshop on Pre-Clovis Sites, Austin, Texas, February 14.

White, J. W. C., and E. J. Steig. 1998. Timing is everything in a game of two hemispheres. *Nature* 394: 717–18.

Williams, C., B. Flower, D. Hastings, and N. Randle. 2008. Meltwater and abrupt climate change in the Gulf of Mexico during the last glacial termination. American Geophysical Union, fall meeting 2008, abstract #PP21C-1442.

Zegura, S. L., T. M. Karafet, L. A. Zhivotovsky, and M. F. Hammer. 2004. High-resolution SNPs and microsatellite haplotypes point to a single, recent entry of Native American Y chromosomes into the Americas. *Molecular Biology and Evolution* 21: 164–75.

PART II

Response

6

Alluvial History and Climate Crises

CLAUDIO VITA-FINZI

INTRODUCTION

One definition of crisis is turning point, as in a fever where a moment of danger is passed. The history of many alluvial valleys includes several such events, when progressive silting gives way to rapid channel incision. The rivers of the Mediterranean Basin and the Near East have long inspired a vigorous debate between those who see the record of successive erosional episodes as primarily a product of climatic factors and those who view it as a narrative of decay for which misguided human activity is largely responsible.

That such interpretative polarization should persist for decades (see Bintliff 2000) says little for the quality of the field data or its analysis. And, whatever the reasons, the continuing uncertainty bears on soil conservation practice, climatic history, and other practical issues. For example, acceptance of erosion as currently inevitable in parts of Eurasia would free funds that are now devoted to hillside terracing and reforestation and divert them to channel regeneration or the construction of check dams designed to trap sediment as well as runoff.

A longstanding instance of the dispute was sparked by field reports of widespread valley silting that postdated the Roman period and that locally attained depths greater than 5 m. The area known to be affected lies between longitudes 10° W and 60° E and latitudes 25° and 45° N. The deposits in question generally consist of well-sorted silt and fine sands that are rich in drab ferrous oxides, contain freshwater molluscs, and have a terminal surface that dips downvalley more steeply than the bed of the original channel and its modern counterpart, attributes inconsistent with the products of accelerated soil erosion and more indicative of perennial or well-distributed channel flow where ephemeral or flashy regimes now prevail. The material that accumulated reflects silt-clay depletion, and thus, the preferential deposition of silt and fine sand; its mineralogy and fossil content indicate localized waterlogging.

An early report referred to these deposits in Greece as the new Elysian Fields (Vita-Finzi 1966a). In Greek mythology, Elysium was part of the Underworld, a final resting place for heroes, carpeted in asphodel rather than cereals,

but above all a notion doubtless colored by the scarcity of fertile bottomlands in the rocky topography of Greece. Silt deposition in medieval times vastly increased the acreage of plains within easy reach of rivers and springs, events that might have robbed the Elysian image of some of its exclusive glamor.

Recent field studies (Casana 2008) confirm that, in the Levant at least, nearly three millennia of sedentary agriculture in the Bronze and Iron Ages had relatively little impact on the land and that, although Roman cultivation and deforestation created the preconditions for soil loss, this had to await the incidence of erosive rainfall after AD 150 at the earliest. In the Orontes Basin, the resulting sediments, locally over 6 m deep, hosted malarial swamps until artificially drained in the 1950s and 1960s. The Arta plain of Greece was similarly rendered productive by twentieth-century drainage. The broadly synchronous timing of aggradation throughout the Old World, with its complex cultural history, in any case is also more consistent with a climatic factor than with a change in agricultural practice or widespread abuse and neglect of the land. These points have been made before (e.g., Hutchinson 1969) but they need to be reiterated because *a priori* opposition to a climatic model (e.g., Fouache 1999) continues to hinder the objective analysis of an important episode in environmental history.

Forty years of spasmodic field research in middle and low latitude Eurasia have raised the number of ^{14}C dates for the valley fill from five in 1969 to about eighty today, a number inadequate to investigate the palaeohydrology of a single headwater stream let alone a complex basin measuring over 2 million km^2 but sufficient to indicate that the onset and progress of silting bore little relation to variations in the progress of settlement. Table 1 supplements an earlier compilation (Vita-Finzi 2008). A full analysis will require information on the depth within the deposit of the dated material and its position along the drainage system; the data to hand merely confirm that deposition was broadly synchronous over a wide range of topographies and lithologies.

The dates are consistent with the simplest explanation for the aggradation, which derives from the work of Leopold (1951) on arroyos in the American southwest, namely, a fall in the relative importance of erosive rains, which resulted from increased cyclonic (relative to convectional) rainfall during a temporary equatorward displacement of the mid-latitude storm tracks. To judge from modeling studies, the requisite displacement of the tropospheric subtropical jet stream could have been prompted by a decrease in solar UV that is indicated by the atmospheric ^{14}C record (Vita-Finzi 2008; Fig. 1).

Aggradation was not everywhere a harmless process confined to uninhabited valley floors. At least three urban sites were partially or wholly overwhelmed by the silt. In Greece, Olympia was excavated in the early twentieth century from beneath the lower terrace of the Alfios; in Algeria, the Roman town of Tubusuctu lies under the deposit in Wadi Soummam; and in Tunisia, "a Roman town" is reported beneath the very low terrace in Wadi es-Sgniffa (Ballais 1995). The sands deposited by the 1966 Florence floods illustrate the speed with which (as Veggiani [1983 and passim] vividly described for medieval northern Italy) urban settlements can be overwhelmed by aggrading rivers.

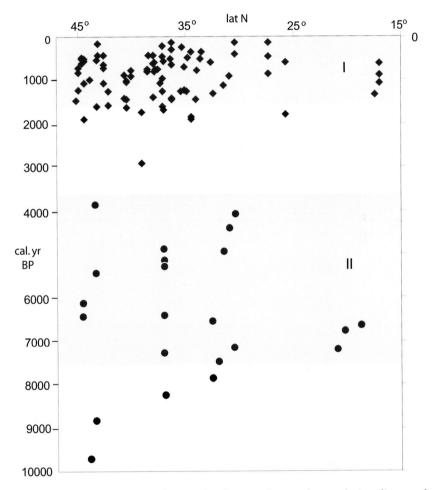

Figure 1. Age/latitude plot of ¹⁴C ages for the two phases of aggradation discussed in the text. Unit I, filled diamonds, Table I, Unit II, filled circles, Table II; error bars omitted for clarity. Shading indicates approximate duration of solar UV minima indicated by atmospheric ¹⁴C after Vita-Finzi (2008).

TABLE 1. Additional ^{14}C ages for Fill I *

Location	Lat.	^{14}C age yr BP	Cal age yr BP§	Lab. no.	Source
ALGERIA					
Chéria-Mezeraa (a)	35°27'	1350±70	1260±65	n.a.	12
ITALY					
Bradano (b)	40°40'	1500±50	1410±65	TO-6860	1
"		980±50	880±60	TO-6859	1
Reno	44°27'	1150±30	1070±55	AA54506	2
"		1150±40	1070±55	AA54513	2
" (a)		1950±40	1910±45	AA54505	2
Sasso	42°36'	1140±50	1065±70	Beta-162902	3
		800±50	735±40	Beta-162903	3
		380±40	420±70	Beta-164915	3
		720±50	655±55	Beta-162901	3
Feccia	43°11'	170±40	150±115	SRR-3253	4
"	"	400±50	425±75	SRR-3252	4
Basento	40°28'	1130±40	1045±55	UGAMS 01911	5
		1553±60	1450±65	LTL 1816A	5
		1090±40	1010±40	UGAMS 01913	5
		1710±40	1630±55	UGAMS 01914	5
Modena	44°29'	520±70	570±55	n.a.	6
"		480±60	520±40	n.a.	6
JORDAN					
W al-Wala (c)	31°31'	1190±50	1120±65	Beta-121656	7
W. Hasa	30°56'	990±120	920±125	Beta-55929	8
MOROCCO					
W Lahouar	30°24'	50±140	135±130	Hv 7053	9
"		370±100	405±90	Hv 7054	9
Rabat	34°02'	1530±80	1440±80	n.a.	10
SAUDI ARABIA					
W. Jizan (d)	17°02'	620±60	610±45	W-4949	11
Jabal al Tirf	17°01'	1140±50	1065±70	W-4944	11
" (a)		980±60	880±65	W-9496	11
W Baysh (a)	17°30'	1430±70	1350±50	B-2978	11
SPAIN					
Fiñana	37°10'	190±120	210±165	Hv 7063	10
"		1695±75	1610±85	Hv 7521	10

*To supplement Vita-Finzi 2008. Dates on charcoal except where shell (a), bone (b), charred material (c), or humic soil (d) is indicated.

§ Ages calibrated using CalPal (quickcal2007ver.1.5), courtesy of the Radiocarbon Laboratory, University of Cologne.

1 Small 1998
2 Eppes et al. 2008
3 Giraudi 2005
4 Hunt and Gilbertson 1995
5 Boenzi et al. 2008
6 Veggiani 1983

7 Cordova 2008
8 Schuldenrein 2007
9 Rohdenburg 1977
10 Gigout 1961
11 Whitney 1983
12 Farrand et al. 1982 in Ballais 1995

The atmospheric ^{14}C curve points to an earlier solar minimum dating from about 7500–4000 years ago, at a time when a fill that locally exceeds a thickness of 15 m accumulated in Jordan, Italy, Tunisia, Morocco, Saudi Arabia, Portugal, and Spain (Table 2).

The explanatory and predictive value of the UV model stands to benefit from confirmation of its validity at similar latitudes in North America. The alluvial history in the southwestern USA (Haynes 1968) includes a depositional unit termed E, which was represented in Nevada, Wyoming, Colorado, Arizona, and New Mexico, and for which the ^{14}C ages then available indicated deposition between ~ 1,600 and 200 cal. yr BP. Subsequently, Cooke and Reeves (1976) found that, despite different initial conditions, arroyo initiation in southern Arizona and coastal California occurred mainly between AD 1850 and 1920, and other studies have further reinforced Haynes' 1968 scheme in Arizona (Waters 1985), Texas and Oklahoma (Hall 1990; Nordt 2004), Nebraska (Daniels and Knox 2005), Montana (Brakenridge 1980), and in the Chihuahua area of Mexico (Nordt 2003). The maximum age for the fill in the Central Plateau of Mexico is 1,880±120 cal yr BP, consistent with the suggestion that deposition began earlier in the south of that country (Vita-Finzi 1977). The second Eurasian fill also finds American counterparts. The corresponding deposit in the scheme of Haynes (1968) is Deposition C (1 and 2), for which he reported ages between 4,470±360 and 7,740±80 cal yr BP. In Washington state, the Clearwater Basin includes terraces dated to ~ 4000–8000 yr BP (Qt5) (Wegmann and Pazzaglia 2002). In Cowhouse Creek, Texas, the FH alluvium, which underlies the main (T1) terrace, yields ages of between 5980±245 and 7710±85 cal yr BP (Nordt 2004).

Of course, the American record has already been attributed in part to changes in the seasonality of small and large rains. In New Mexico, aggradation is linked to times when summer thunderstorms are less frequent, and large floods are correspondingly rare (Mann and Meltzer 2007); synchronous episodes of stream incision during the Holocene from many parts of the northern mid-latitudes (Brakenridge 1980), and of flood frequency in the Upper Mississippi Valley and the American southwest (Ely et al. 1993; Knox 2000), are likewise set in the context of changes in atmospheric circulation patterns. Knox (2000) argued that, in the early twentieth century, tropospheric westerly circulation patterns over the Midwest had a relatively strong meridional component during years with large floods and a relatively strong zonal component during years of small floods. In the Sierra Nevada, droughts in AD 890–1,110 and 1,210–1,350 are thought to have resulted from a reorientation of the mid-latitude storm tracks (Stine 1994). In the northern prairies, lake sediment cores likewise indicate changes in the shape and location of the jet stream and associated storm tracks during the last 2,000 years (Laird et al. 2003). The present study merely provides a feasible explanation for the recurrence of a seasonal pattern favouring aggradation throughout the low-middle latitudes of the northern hemisphere.

TABLE 2. ^{14}C ages for Fill II

Location	Lat.	^{14}C age yr BP	Cal age yr BP§	Lab. no.	Source
ITALY					
Bidente	43°59′	8700±70	9710±115	AA-64959	1
Tenna	43°24′	3570±70	3865±100	Beta-78660	1
Potenza	43°22′	4680±100	5415±125	n/a	1
Reno	44°24′	5330±40	6110±70	AA-61344	1
"		5630±40	6405±50	AA-61346	1
Musone (a)	43°26′	7970±50	8840±110	AA-64963	1
JORDAN					
W Hasa	30°56′	3950±150	4415±225	Q-729	2
W al Wala	31°31′	4350±60	4950±75	Beta-133945	3
W ash-Shallalah (c)	32°36′	5740±50	6550±70	Beta-147282	3
MOROCCO					
Tarhazoute	30°32′	6295±180	7175±200	HV-7055	4
W. Lahouar	30°24′	3715±130	4090±185	HV-7052	4
PORTUGAL					
Castelejo	37°06′	7450±90	8270±85	Beta-2908	5
SAUDI ARABIA					
Hamda 1	8°44′	5700±250	6535±280	W-4112	6
Mokhyat	20°17′	5930±300	6790±340	W-3956	6
Hamdah	19°00′	6120±110	7005±145	B-3462	6
W. al Miyah	20°57′	6350±350	7180±360	W-3282	6
SPAIN					
Guadalentín	37°05′	4305±55	4905±55	Ly-7151	7
"		4500±100	5140±150	SUA-2040	7
"		4520±90	5160±140	SUA-2037 7	
"		4610±150	5275±210	SUA-2038	7
"		5610±330	6430±380	SUA-2039	7
"		6340±60	7280±75	Ly-229/OxA	7
TUNISIA					
Bir al Amir (a)	32°30′	7026±175	7865±160	C 3541	8
Jenain (a)	32°00′	6580±350	7495±45	Gif 8474	8

Dates on charcoal except where shell (a) or "charred material" (c) is indicated.

§ Ages calibrated using CalPal (quickcal2007ver.1.5), courtesy of the Radiocarbon Laboratory, University of Cologne.

1 Wegmann and Pazzaglia 2009
2 Vita-Finzi 1966
3 Cordova 2008
4 Rohdenburg 1977
5 Devereux 1983
6 Whitney 1983
7 Calmel-Avila 2000
8 Ballais 1995

Both alluvial units represent conditions favourable for floodwater farming. Indeed, the older deposit in Wadi Hasa, Jordan, was used to illustrate the notion of "geological opportunism," where a population is neither cosseted nor cowed by the environment but exploits the prevailing conditions. Opportunism was proposed to counter the two extreme positions—possibilism and determinism—then occupied by some historical geographers (Vita-Finzi 1969). The mild, repetitive flooding indicated by its finely stratified deposits is reminiscent of the hydrological regime that prevailed during Hopi occupation of New Mexico, when the economy depended heavily on floodwater farming (Bryan 1929) until arroyo formation in the twentieth century put paid to the opportunitiy.

In Wadi Hasa, the sediments flooring the main valley floor yield artifacts that date from the middle Neolithic, with the implication that deposition began during or soon after 8000 yr BP. Charcoal at a depth of 1 m was dated to 4415±225 cal yr BP, and a Roman wall within a channel cut into the deposit shows that aggradation had ended 2,000 years ago (Copeland and Vita-Finzi 1978). According to Schuldenrein (2007), the fill is the morphostratigraphic equivalent of the valley floors in many other parts of the Near East, and their age is therefore critical to understanding the origins of Near Eastern subsistence strategies. In nearby Wadi al-Wala, the Iskanderite Alluvium underlay the floodplain during Chalcolithic and Early Bronze Age occupation of the site of Khirbet Iskander. In Wadi ash-Shallalah, deposition of the related deposit (here consisting of Units I and II) coincided with Chalcolithic and Early Bronze Age occupation of Zeiraqoun (Cordova 2008).

In parts of the Maghreb, a similar deposit exposed by 3–15 m was laid down in 8300–4000 cal yr BP; its composition and fauna indicate greater humidity than now locally prevails (Ballais 1995; Benito 2003). In the lower reaches of the Bradano, Basento, and Cavone, in southern Italy, a major terrace (T2) is underlain by 15 m of deposits indicative of rapid alluviation and dating from about the same time (Abbott and Valastro 1995). Further north, in the Apennines of Romagna and the Marche, a fill (t8) with a maximum elevation of 11 m above the channel ranges in age from 3865±95 to 6405±45 (Wegmann and Pazzaglia 2009). The corresponding deposit in the Gudalentín valley of SE Spain consists of silts emplaced by successive overbank floods ("l'abondance des limons de débordement qui la composent fait penser à une mise en place par crues successives" [Calmel-Avila, 2000, 154]).

In the Jordanian wadis, and doubtless elsewhere, stream incision eventually ruled out flood irrigation. In the view of Cordova (2008), this was possibly why many Early Bronze Age sites near the floodplains were not reoccupied. Nevertheless trenching was beneficial where waterlogging had hindered mobility and promoted malaria, even if irrigated agriculture might now call for damming or pumping. In the Rhone-Alpes region of France malaria was endemic in the nineteenth century; in the Isère valley the drainage of marshes and the elimination of braided river channels reduced mortality by 60 percent between 1870 and 1900 (Sérandour et al. 2007).

DISCUSSION

At least three more millennial-scale UV plateaux (i.e., solar minima) are displayed by the atmospheric ^{14}C record for 0–25,000 cal yr BP. Their interpretation is especially delicate as they occur within what is conventionally considered the last glacial period and, like the two phases considered earlier, they are underlain by a massive secular decline in ^{14}C whose origin is uncertain. If river behavior in Eurasia was indeed affected, we may have stumbled on an explanation for the multiplicity of river terraces displayed by many major rivers (as with the thirteen or more successive alluviation episodes that have been identified for the last 200,000 years in the Mediterranean) that were broadly synchronous across the region (Macklin et al. 2002).

The minor fluctuations in the curve may prove equally informative. Thus, just as what appeared to be noise in ^{18}O data from deep-sea cores is now interpreted as a series of short-lived melting episodes ("Heinrich events") during the last glacial period, so may short-term fluctuations throughout the available $\Delta^{14}C$ archives emerge as the creature of solar-terrestrial interaction, with meteorological consequences yet to be identified. The Maunder Minimum and other century-scale anomalies for which there is abundant historical (and sunspot) corroboration emerge as puny when compared with the millennial phases, and commonplace once the frequency of comparable ^{14}C anomalies is grasped, but their hydrological significance could prove historically profound.

Fluctuations in solar activity have long been invoked to explain facets of human history. Consider, in particular, the work of Ellsworth Huntington (e.g., 1915) on the impact of climatic changes on civilization and the emphasis he placed in his analysis on solar luminosity, shifting storm tracks, and the incidence of cyclonic rain. The significance of alluvial settings in human history is obvious, but in many recent geoarchaeological studies it is eclipsed by ruminations on specific components of the climate itself, notably rainfall, rather than the fluvial response.

By the same token the economic and social impact of historical aggradation tends to be neglected in favor of analyses of drought, flooding, and other components of the precipitation regime. Granted their importance in determining success and failure in agriculture, urban life, navigation, and so forth, these items are on the whole more transitory than soil creation and destruction and by the same token more difficult to extract from the field evidence.

If the link between stream behavior and solar UV can be corroborated, the study of the social impact of climatic crises will benefit in three key ways. First, there will now be a geographical framework incorporating latitudinal gradients with which to compare archaeological and historical evidence, seek novel data, and perhaps explain puzzling gaps and anomalies. Second, the rough periodicity of the ^{14}C and ^{10}Be peaks and troughs points to several more periods than the two discussed here, when rivers in the area under review silted up and then cut down, with all the attendant benefits and drawbacks. Third, one might expect to find evidence for parallels to these events in the Southern Hemisphere, and different (though not necessarily contrary)

conditions in the zones north and south of the aggrading northern latitudes. In short, there will be predictive environmental models to be tested by the discoveries of archaeology and palaeoanthropology, a reversal of roles that cannot fail to stimulate the field scientist even if the answers may have to be substantially deferred.

REFERENCES

Abbott, J. T., and S. Valastro Jr. 1995. The Holocene alluvial records of the chorai of Metapontum, Basilicata, and Croton, Calabria, Italy. In *Mediterranean Quaternary river environments*, ed. J. Lewin, M. G. Macklin, and J. C. Woodward, 195–205. Rotterdam: Balkema.

Ballais, J.-L. 1995. Alluvial Holocene terraces in eastern Maghreb: climate and anthropogenic controls. In *Mediterranean Quaternary river environments*, ed. J. Lewin, M. G. Macklin, and J. C. Woodward, 183–94. Rotterdam: Balkema.

Benito, G. 2003. Palaeohydrological changes in the Mediterranean region during the late Quaternary. In *Palaeohydrology*, ed. K. J. Gregory and G. Benito, 125–36. New York: Wiley.

Bintliff, J. 2000. Landscape change in Classical Greece: a review. In *Geoarchaeology of the landscapes of Classical Antiquity*, ed. F. Vermeulen and M. de Dapper, 49–70. Leiden, Babesch.

Brakenridge, G. R. 1980. Widespread episodes of stream erosion during the Holocene and their climatic cause. *Nature* 283: 655–56.

Bryan, K. 1929. Floodwater farming. *Geographical Review* 19: 444–56.

Calmel-Avila, M. 2000. A study of Holocene palaeoenvironments in the Lower-Guadalentín basin (Murcia, Spain). *Géomorphologie* 3: 147–60.

Casana, J. 2008. Mediterranean valleys revisited: linking soil erosion, land use and climate variability in the Northern Levant. *Geomorphology* 101: 429–42.

Cooke, R. U., and R. W. Reeves. 1976. *Arroyos and environmental change in the American south-west*. Oxford: Clarendon.

Copeland, L., and C. Vita-Finzi. 1978. Archaeological dating of geological deposits in Jordan. *Levant* 10: 10–25.

Cordova, C. E. 2008. Flood-plain degradation and settlement history in Wadi al-Wala and Wadi ash-Shallalah, Jordan. *Geomorphology* 101: 443–57.

Daniels, J. M., and J. C. Knox. 2005. Alluvial stratigraphic evidence for channel incision during the Mediaeval Warm Period on the central Great Plains, USA. *The Holocene* 15: 736–47.

Devereux, C. M. 1983. *Recent erosion and sedimentation in southern Portugal.* Ph.D. diss., University of London.

Ely, L. L., Y. Enzel, V. R. Baker, and D. R. Cayan. 1993. A 5000-year record of extreme floods and climate change in the southwestern United States. *Science* 262: 410–12.

Eppes, M., R. Bierma, D. Vinson, and F. Pazzaglia. 2008. A soil chronosequence study of the Reno River Valley, Italy. *Geoderma*, doi:10.1016/j.geoderma. 2008.07.011.

Farrand, W. R., C. H. Stearns, and H. E. Jackson. 1982. Environmental setting of Capsian and related occupations in the high plains of eastern Algeria. *Bulletin of the Geological Society of America* 14: 487.

Fouache, E. 1999. L'alluvionnement historique en Grèce occidentale et au Peleponnèse. *Bulletin de Correspondance Hellénique, Supplément* 35. Athens.

Gigout, M. 1961. Vérification de la datation de deux dépôts quaternaries de Rabat (Maroc). *Comptes Rendus de la Societé Géologique de France* 8: 228.

Giraudi, C. 2005. Late-Holocene alluvial events in the Central Apennines, Italy. *The Holocene* 15: 768–73.

Hall, S. A. 1990 Channel trenching and climatic change in the southern U.S. Great Plains. *Geology* 18: 342–45.

Haynes, C .V., Jr. 1968. Geochronology of late-Quaternary alluvium. In *Means of correlation of Quaternary successions,* ed. R. B. Morrison and H. E. Wright Jr., 591–631. Salt Lake City: University of Utah Press.

Hunt, C.O., and D. D. Gilbertson. 1995. Human activity, landscape change and valley alluviation in the Feccia valley, Tuscany, Italy. In *Mediterranean Quaternary river environments,* ed. J. Lewin, M. G. Macklin, and J. C. Woodward, 167–76. Rotterdam: Balkema.

Huntington, E. 1915. *Civilization and climate.* New Haven: Yale University Press.

Hutchinson, J. 1969. Erosion and land use: the influence of agriculture on the Epirus region of Greece. *Agricultural History Review* 17: 85–90.

Knox, J. C. 2000. Sensitivity of modern and Holocene floods to climate change. *Quaternary Science Review* 19: 439–57.

Laird, K. R., B. F. Cumming, S. Wunsman, J. A. Rusak, R. J. Oglesby, S. C. Fritz, and P. R. Leavitt. 2003. Lake sediments record large-scale shifts in moisture regimes across the northern prairies of North America during the past two millennia. *Proceedings of the National Academy of Sciences* 100: 2483–88.

Leopold, L. B. 1951. Rainfall frequency: an aspect of climatic variation. *Transactions of the American Geophysical Union* 32: 347–57.

Macklin, M. G., I. C. Fuller, J. Lewin, G. S. Maas, D. G. Passmore, J. Rose, J. C. Woodward, S. Black, R. H. B. Hamlin, and J. S. Rowan. 2002. Correlation of fluvial sequences in the Mediterranean basin over the last 200 ka and their relationship to climate change. *Quaternary Science Reviews* 21: 1633–41.

Mann, D. H., and D. J. Meltzer. 2007. Millennial-scale dynamics of valley fills over the past 12,000 ^{14}C yr in northeastern New Mexico, USA. *Bulletin of the Geological Society of America* 119: 1433–48.

Nordt, L. 2003. Late Quaternary fluvial landscape evolution in the desert grasslands of northern Chihuahua, Mexico. *Bulletin of the Geological Society of America* 115: 596–606.

———. 2004. Late Quaternary alluvial stratigraphy of a low-order tributary in central Texas: a response to changing climate and sediment supply. *Quaternary Research* 62: 289–300.

Reimer, P. J., M. G. L. Baillie, E. Bard, A. Bayliss, J. W. Beck, C. J. H. Bertrand, P. G. Blackwell, C. E. Buck, G. S. Burr, K. B. Cutler, P. E. Damon, R. L. Edwards, R. G. Fairbanks, M. Friedrich, T. P. Guilderson, A. G. Hogg, K. A. Hughen, B. Kromer, G. McCormac, S. Manning, C. B. Ramsey, R. W. Reimer, S. Remmele, J. R. Southon, M. Stuiver, S. Talamo, F. W. Taylor, J. van der

Plicht, and C. E. Weyhenmeyer. 2004. IntCal04 terrestrial radiocarbon age calibration, 0-26 cal kyr BP. *Radiocarbon* 46: 1029–58.

Rohdenburg, H. 1977. Neue ^{14}C-daten aus Marokko und Spanien und ihre Aussagen for die Relief- und Bodenwicklung im Holozän und Jungpleistiozän. *Catena* 4: 215–28.

Schuldenrein, J. 2007. A reassessment of the Holocene stratigraphy of the Wadi Hasa terrace and Hasa formation, Jordan. *Geoarchaeology* 22: 559–88.

Sérandour, J., J. Girel, S. Boyer, P. Ravanel, G. Lemperière, and M. Raveton. 2007. How human practices have affected vector-borne diseases in the past: a study of malaria transmission in Alpine valleys. *Malaria Journal* 6, 115; doi:10.1186/1475-2875-6-115.

Small, A., C. Small, I. Campbell, M. Mackinnon, T. Prowse, and C. Sipe. 1998. Field survey in the Basentello valley on the Basilicata-Puglia border. *Classical Views* 42: 337–71.

Stine, S. 1994. Extreme and persistent drought in California and Patagonia during mediaeval time. *Nature* 369: 546–49.

Veggiani, A. 1983. Degrado ambientale e dissesti idrogeologici indotti dal deterioramento climatico nell'alto medioevo in Italia. I casi riminesi. *Studi Romagnoli* 34: 123–46.

Vita-Finzi, C. 1966a. The new Elysian Fields. *American Journal of Archaeology* 70: 175–78.

———. 1966b. The Hasa Formation. *Man* 1: 386–90.

———. 1969. Geological opportunism. In *The Domestication and exploitation of plants and animals*, ed. P. J. Ucko and G. W. Dimbleby, 31–34. London: Duckworth.

———. 1977. Quaternary deposits in the central plateau of Mexico. *Geologische Rundschau (International Journal of Earth Science)* 66: 99–120.

———. 2008. Fluvial solar signals. In *Landscape evolution, denudation, climate and tectonics over different time and space scales*, ed. K. Gallagher, S. Jones, and J. Wainwright, 105–15. London: Geological Society Special Publication 296.

Waters, M. R. 1985. Late Quaternary stratigraphy of Whitewater Draw, Arizona: implications for regional correlation of fluvial deposits in the American Southwest. *Geology* 13: 705–8.

Wegmann, K. W., and F. J. Pazzaglia. 2002. Holocene strath terraces, climate change, and active tectonics: the Clearwater River basin, Olympic Peninsula, Washington State. *Bulletin of the Geological Society of America* 114: 731–44.

———. 2009. Late Quaternary fluvial terraces of the Romagna and Marche Apennines, Italy: climatic, lithologic, and tectonic controls on terrace genesis in an active orogen. *Quaternary Science Reviews* 28: 137–65.

Whitney, J. W. 1983. *Erosional history and surficial geology of western Saudi Arabia*. Technical Record USGS-TR-04-1 Ministry of Petroleum and Mineral Resources, Jiddah.

7

Pleistocene Climate Change and Human Occupation of Eastern North Africa

JENNIFER R. SMITH

Any evaluation of the role of abrupt climate change or climate events in the course of human history requires, ideally, a detailed record both of human behavior or activity across the time of the event, and of the nature, timing, and magnitude of the event itself. As one moves deeper into the archaeological and geologic records, achieving the level of detail required to evaluate hypotheses regarding causes of change in human activity becomes more difficult. Once we begin to investigate these questions in a time before texts, before significant architecture or complex material cultures, when lithic artifacts represent the only available cultural record, demonstrating a complicated behavioral response to an environmental change is exceedingly challenging, if not impossible.

One way to address these challenges is to simplify the questions being asked, commensurate with the nature of the data available. The presence or absence of humans in a region can be taken as a simple index of response to climate change if (1) climate events are of sufficient magnitude, or (2) a human community persists close to a threshold of habitability, where a slight change in climate will make the local environment habitable or uninhabitable due to resource availability. The nature of that human response can be characterized by determining whether or how (in what spatial or temporal pattern) a previously unoccupied region is exploited as a place to settle once it becomes viable, or a place through which to migrate. North Africa during Pleistocene time is an ideal setting for investigation of this type of question, since much of the landmass has oscillated between habitable and uninhabitable due to changes in magnitude and/or timing of rainfall.

Two principal datasets are required to evaluate the effects of climate change on hominins in North Africa: an environmental sequence (to document the nature and magnitude of climatic variation) and an archaeological sequence (to indicate locations of occupation). The successive periods of arid and humid conditions that make Pleistocene North Africa so well suited conceptually as a place to investigate climate's influence on hominins present challenges to the development of the necessary datasets. Sediments

deposited during humid phases in springs, lakes, wetlands, or rivers are easily eroded by the wind (deflated) during subsequent arid phases (see, e.g., Churcher and Kleindienst *in press*; Smith et al. 2004a, b). Thus, terrestrial archives of North African paleoclimate are frequently discontinuous, and it is particularly the sediments recording the transitions from humid to arid climates, deposited at the end of humid phases, that are most likely to be lost to erosion. This systemic discontinuity impedes any potential direct reconstruction of human response to the aridification of North Africa. Marine cores from the Red Sea, Mediterranean Sea, or the Atlantic Ocean can provide more continuous climate sequences for the last several hundred thousand years (e.g., Hemleben et al. 1996; Larrasoaña et al. 2003; Legge et al. 2008; Moreno et al. 2001; Stein et al. 2007; Weldeab et al. 2007); however, these cores often integrate climate signals over large regions and cannot typically be used to reconstruct environmental variables at a scale necessary to constrain resource availability for local human populations. The archaeological record is also negatively impacted by eolian erosion; thus, the vast majority (although by no means all) of archaeological materials across the Sahara are found in surface contexts (e.g., Olszewski et al. 2005; Hawkins 2001; Kleindienst 1999). Either artifacts were primarily deposited on stable, nondepositional surfaces, such as desert pavements, or they were deposited within accumulating sediments, which were later deflated, leaving the artifacts as lags. In both circumstances, chronological control on artifact deposition, or a reconstruction of the environmental context of occupation, is extremely difficult to achieve.

The localities where Pleistocene archaeological materials are found in directly datable contexts, then, are particularly valuable for developing a coupled climate/culture narrative. Some of these locations, however, are in regions likely to have experienced environmental histories different from that of the Sahara proper. Rock shelters and caves of Morocco (e.g., Smuggler's Cave, Dar es-Soltan, Rhafas) and Libya (Haua Fteah) are notable for their stratigraphic sequences (e.g., McBurney 1967; Barton et al. 2009; Mercier et al. 2007; Nespoulet et al. 2008); however, these locations are generally within the narrow coastal belt currently affected by Mediterranean-sourced winter rainfall (see below), and as such are likely to have experienced climate chronologies different from the bulk of inland North Africa. The consistent availability of marine resources to coastal dwellers would have made subsistence strategies different from those adopted in inland regions; thus, the nature of hominin response to environmental change is also likely to have been different along the coasts. Archaeological materials are also found in stratigraphic context within Nile alluvial deposits (Wendorf and Schild 1976; Giegengack 1968; Vermeersch et al. 1982; Vermeersch 2000; Van Peer et al. 2003; Van Peer 1998); however, the Nile, with its vast watershed that drains several climate regimes, would have been largely buffered from climate variation due to local rainfall fluctuations over the Sahara. Both the Mediterranean coast and the Nile Valley are likely to have represented refuges, to which hominins migrated during times when the rest of North Africa was uninhabitable. Thus, these contexts will not be discussed further here, as the influence of climate

on hominins would be much more difficult to discern there than in regions that became at times uninhabitable. Although the groundwater-fed Saharan oases—for example, those of the Western Desert of Egypt—would also have provided a refuge during conditions like those of today (Kleindienst 1999), the dramatic fluctuation in the availability of water resources (from relatively limited spring sources to large lakes) due to climate variability would certainly have affected occupants of the region; thus environmental and archaeological evidence from oases will be considered here.

MODERN CLIMATE OF NORTH AFRICA

In order to understand the effects of climate variation on North African populations, it is necessary first to describe modern climates in North Africa as a baseline with which to compare past conditions. Under the modern climate regime, habitable regions of North Africa are generally confined to the Mediterranean, Atlantic, and Red Sea coasts; the Nile Valley and Delta; and the desert oases. Winter rainfall occurs along the Mediterranean margin of North Africa (Issar 2003), and can occasionally penetrate as far south as 20° N (Fig. 1; Geb 2000). Summer monsoonal rains water the southern margin of the Sahara, and under modern atmospheric conditions may result in storms as far north as 32° N (Geb 2000). Mean annual rainfall across the Northern Sahara ranges from 1 to 689 mm/yr, with higher precipitation found in the west (Warner 2004).

Humid-phase Environments of Pleistocene North Africa

The principal climate "crises" affecting hominin occupants of North Africa would have been transitions from relatively humid conditions to arid

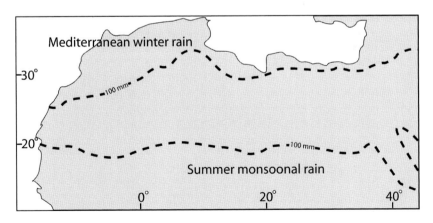

Figure 1. Modern rainfall regimes, as indicated by the 100 mm isohyet over North Africa. Modified from Arz et al. 2003.

climates, similar to the modern conditions described above. The severity of each crisis would have been related to the magnitude of the change in rainfall amounts and the rapidity with which the change took place. Rates of change are often difficult, if not impossible, to constrain, given that transitional sediments are frequently (although not always; e.g., McKenzie 1993) removed by erosion. These same sedimentary archives, however, are well suited to reconstructing humid-phase landscapes. In many cases, the sedimentology and geochemistry of humid-phase sediments may also be used to impose constraints on rainfall amounts.

Lakes

Perhaps the most dramatic evidence for a significant departure from modern Saharan conditions in the recent geologic past is represented by the extensive Pleistocene lacustrine deposits found in several locations across North Africa. Lake sediments have been described from the oasis basins of central Egypt (Kieniewicz and Smith 2009; Churcher, Kleindienst, and Schwarcz 1999; Churcher and Kleindienst, in press), southern Egypt/northern Sudan (Wendorf, Schild, and Close 1993; Szabo, Haynes, and Maxwell 1995; Schwarcz and Morawska 1998; McKenzie 1993; Hill 2001), and western Libya (Geyh and Thiedig 2008; Drake et al. 2008; Armitage et al. 2007). The Tunisian Great Chotts are also known to have expanded periodically, and their water to have become less saline (Causse et al. 2003, 1989). Reconstructions of the surface area of the central Egyptian and Libyan paleolakes based on the distribution of lacustrine sediments and their surrounding topography indicate that during Marine Isotope Stage (MIS) 5 (~130,000–75,000 years), these lakes covered ~1400–1700 km^2 each (Kieniewicz and Smith 2009; Geyh and Thiedig 2008; Drake et al. 2008; Smith 2010).

The nature of the resources and habitats offered by these lakes during humid phases would have varied somewhat, both spatially and temporally. Significant water bodies are indicated in each of the above regions during at least portions of MIS 5, but some basins contain evidence of lake highstands both prior to and subsequent to MIS 5 (Wendorf, Schild, and Close 1993; Geyh and Thiedig 2008; Drake et al. 2008). Lake volumes and salinity would have varied, in some cases substantially, over the life span of each lake (Zouari, Chkir, and Causse 1998; Wendorf, Schild, and Close 1993; Geyh and Thiedig 2008), although at least some lake sediments indicate perennial freshwater conditions lasting for millennia (McKenzie 1993; Kieniewicz and Smith 2007, 2009; Churcher and Kleindienst in press). This intrahumid-phase environmental variability would have caused stresses on occupants of the region during times of shrinking and increasingly saline lakes; thus, subsistence strategies may have changed over the course of a single humid phase.

Lacustrine environments that prevailed during Saharan humid phases would have been spatially variable as well. The sedimentology of Egyptian lacustrine deposits indicates that these lakes were not merely single deep basins with shallow margins; they consisted of discrete sub-basins, iso-

lated from each other at shallower water depths (Wendorf, Schild, and Close 1993; Churcher and Kleindienst in press). Vegetated shallow-water facies are common in these lakes (see, e.g., Kieniewicz and Smith 2009), suggesting that marshy environments were frequently extensive. Vertebrate fossils recovered from lacustrine sediments in southern and central Egypt are consistent with a savanna fauna utilizing the lakes as a water source (Wendorf, Schild, and Close 1993; Churcher, Kleindienst, and Schwarcz 1999; Churcher and Kleindienst in press). Thus, game and water would have been plentiful during the heights of humid phases. However, during lake lowstands much of the lake basin that was formerly open water would have been covered by marsh. Large swampy areas may have represented barriers to through-going travel in the region.

Springs

Spring deposits are common throughout North Africa. These deposits indicate that during humid phases, spring-dominated landscapes may have consisted simply of small isolated pools at individual spring vents, or may have been characterized by relatively high-discharge springs that supported rivers or lakes, depending on the geomorphic setting (Smith and Giegengack 2003). Most spring deposits described in North Africa are tufas or travertines, authigenic, calcium-carbonate-rich sediments precipitated from groundwaters sourced in limestone aquifers (e.g., Akdim and Julia 2005; Boudad et al. 2003; Brook et al. 2003; Crombie et al. 1997; Hamdan 2000; Nicoll, Giegengack, and Kleindienst 1999; Smith, Giegengack, and Schwarcz 2004; Smith et al. 2004a, b). However, springs sourced in the Nubian Aquifer, which underlies much of the Eastern Sahara (Thorweihe 1990; Sturchio et al. 2004), may precipitate iron-rich sediments (Kleindienst 1999; Caton-Thompson 1952; Brookes 1993; Adelsberger and Smith in review). In most locations, spring deposits record times of enhanced spring discharge relative to the present; this enhanced discharge is presumed to have been caused by increased regional rainfall. However, in Morocco, the timing of travertine formation from springs located on active faults may be related to tectonic activity rather than to climate change; thus, these sediments present additional challenges to paleoenvironmental interpretation (Wengler et al. 2002; Weisrock et al. 2006; Boudad et al. 2006, 2003; Akdim and Julia 2006, 2005).

Spring-fed environments would have varied substantially over the course of a humid phase. For instance, spring mounds in the Dakhleh Oasis basin would have been exposed as islands when the level of the surrounding lake dropped low enough, but springs would have discharged directly into the lake when lake levels were high. During drier times, when rainfall was not sufficient to allow for large lakes, spring discharge would have supported wetlands. The presence of reed casts, algal/bacterial mats, and iron-sulfate minerals such as jarosite in predominantly iron oxyhydroxide sediments in the Dakhleh Oasis spring mounds suggests acidic bog environments of deposition (Adelsberger and Smith in review).

Case Study: Spring Tufas of Kharga Oasis, Egypt

The fossil-spring tufas of Kharga Oasis (Fig. 2) are a historically significant example of Pleistocene humid-phase deposits containing evidence of contemporaneous hominin occupation. The tufas were first studied systematically by Gertrude Caton-Thompson and Elinor Gardiner in the 1930s (Caton-Thompson 1952). It was these spring deposits and their associated artifacts on which Caton-Thompson based her regional archaeological stratigraphy. The spring tufas and interbedded lacustrine, paludal, and fluvial sediments along the north-south trending Libyan Escarpment bordering Kharga Oasis preserve a record of small ponds, rivers, and marshy areas at elevations above the floor of the oasis depression, where the more extensive humid-phase lakes would have been found (Sultan et al. 1997; Smith et al. 2007, 2004a, b; Nicoll, Giegengack, and Kleindienst 1999; Kleindienst et al. 2008; Kieniewicz and Smith 2007; Caton-Thompson 1952; Brook et al. 2003). Middle and Early Stone Age lithic artifacts have been found in both primary and secondary (i.e., reworked) contexts within these sediments; however, late Middle and Late Stone Age archaeological materials have generally only been found in surface contexts, or within clastic playa sediments filling tufa basins, not incorporated within or overlain by tufa (see, e.g., Smith et al. 2007; Hawkins et al. 2001; Kleindienst et al. 2008; Caton-Thompson 1952).

The environments of Early and Middle Stone Age occupation or utilization of the escarpment region, as recorded by spring sediments, would have been spatially heterogeneous, with tufa barriers impounding small (several m^2) ponds and wetlands, connected by channels (Smith et al. 2004a, b; Nicoll, Giegengack, and Kleindienst 1999; Crombie et al. 1997). Aquatic vegetation, particularly reeds, would have been common, and fig and palm trees grew along the spring-fed rivers (Caton-Thompson 1952; Smith et al. 2004a, b). The geochemistry and stratigraphic extent of lacustrine sediments interbedded with tufas at Wadi Midauwara (see Fig. 2) indicate that spring discharges during the MIS 5 humid phase were high enough to support a small, perennial, freshwater lake for at least several millennia (Smith, Giegengack, and Schwarcz 2004; Smith et al. 2007; Kieniewicz and Smith 2007). Presumably these springs, rivers, and ponds would have attracted game, although no fossil evidence of such has been found in Kharga. Lithic raw material would also have been available from chert-bearing limestones comprising the upper portions of the escarpment (Hawkins et al. 2001; Kleindienst et al. 2008). Thus, these spring-fed areas along the Kharga Escarpment would have been excellent habitat for hominins, and the drying of these regions a significant crisis for populations exploiting the resources therein.

During "lesser" humid phases, however, such as those following MIS 5 (e.g., Smith in press), water-resource availability would probably have been substantially different. Based on the near-total absence of spring carbonates dating to these humid phases, spring flow was likely highly localized, or surface water generated mainly from runoff, with little to no groundwater activity along the Escarpment (Caton-Thompson 1952; Smith et al. 2004a,b). Archaeological materials, and the occasional direct date on calcretes

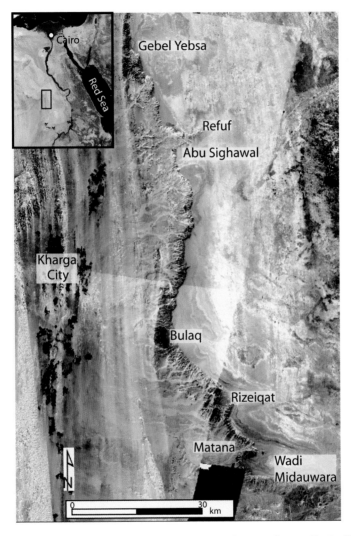

Figure 2. Locations of Pleistocene fossil-spring tufas in Kharga Oasis, Western Desert, Egypt (Smith, in press).

(Churcher, Kleindienst, and Schwarcz 1999), or remobilized uranium (Osmond and Dabous 2004) from the Western Desert, provide the only direct evidence for mildly enhanced water availability relative to today. If the rainfall was brought either by northward migration of the summer monsoon, or by southward penetration of winter Mediterranean rains, water resources would likely only have been seasonal, since it does not seem that regional aquifers were significantly charged during these times. The "crises" affecting occupants of the Western Desert during these minor humid phases would have been less severe than the crises at the ends of the more substantial climatic variations, but would still have necessitated a response. Populations would have had to either move into oasis lowlands, and subsist on more limited water supplies from long-lived springs sourced in the Nubian Aquifer, or migrate into the Nile Valley.

Rivers

Drainage networks that developed during times of significant rainfall in the Sahara may be difficult to date precisely, or may have been obscured by arid-phase eolian deposition. Arguments for the existence of extensive surficial drainage networks in the Sahara have been made based on radar imagery, Mediterranean sediment chemistry, geomorphology, and archaeology (see, e.g., Rohling et al. 2002; McHugh, McCauley et al. 1988; McHugh, Breed et al. 1988; McCauley et al. 1982; Haynes et al. 1997; Drake et al. 2008). Drainage networks were most likely established during the Tertiary Period; in Pleistocene time, the rivers of southern Egypt would most likely have been small, braided streams within broader incised valleys, with floodplain marshes and small lakes during times of high river discharge or frequent flooding (McHugh, McCauley et al. 1988; Drake et al. 2008). Direct dating of these features is extremely difficult. Major drainage systems across Libya are believed to have been active at some point during the Pleistocene, certainly around 120 ka, but the number of activity phases and their timing cannot be further constrained (Paillou et al. 2009; Osborne et al. 2008; Drake et al. 2008).

Timing of Humid Phases

More thorough discussions of the cyclicity and intra-regional variability of humid-phase timing in North Africa can be found in Smith, Kieniewicz, and Asmerom (submitted) and Smith (in press); a brief recapitulation of those studies is presented here. The errors associated with direct dates on humid-phase sediments (generally, 1,000–10,000 years) make it difficult to establish discrete ages for the onset and end of a humid period. Nonetheless, the frequency and distribution of ages on humid-phase sediments over the last 300,000 years indicate that humid phases were not simply associated with interglacial periods; they occurred with a periodicity that would be expected if orbital precession were the principal control on Saharan rainfall.

When timing of humid phases is compared across geographic subdivisions of North Africa (Mediterranean North Africa, Northern Sahara, and

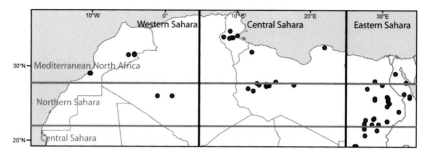

Figure 3. Locations of paleoenvironmental records from North Africa, modified from Smith, in press.

Central Sahara; East, Central, and West Sahara), some significant differences can be observed (Smith in press). The statistical comparison of climate records across these regions is hindered by varying sample sizes; far more paleoenvironmental research has been carried out in the Eastern and Northern Sahara than in the other regions (Fig. 3). Qualitatively, however, humid phases later than MIS 5 are better represented in the Western Sahara/Mediterranean North Africa than they are in the other regions, where MIS 5 seems to be the wettest time in the last 350,000 years (Smith in press). Most of the paleoenvironmental records from the Western Sahara/Mediterranean North Africa are within the modern "Mediterranean" climate zone (see Fig. 1); it may be that winter rainfall was the predominant source of rainfall during the post-MIS 5 humid periods, and therefore these phases were best expressed along the Mediterranean margin of North Africa.

Although humid-phase sediments dated to MIS 5 do dominate paleoenvironmental records in the central, southern, and eastern Sahara, the timing of "peak" humid events, as indicated by maxima in the probability-density distribution of dates on humid-phase sediments from these regions, varies slightly. It is probable that there were wetter (MIS 5a, c, and e) and drier (MIS 5b and d) times within MIS 5; there is enough regional variation in chronology that it appears humid conditions could be found somewhere in North Africa throughout all of MIS 5. This spatial variation in climate conditions would have allowed hominin occupants of the region to adapt to climate variation within MIS 5 by migration within the Sahara, rather than by leaving the region altogether. Habitation would have become significantly more difficult following MIS 5, although at least two relatively minor humid phases did occur around 45–35 and 30–20 ka (Smith, Kieniewicz, and Asmerom submitted), prior to the well-recognized Holocene humid event (see, e.g., Hoelzmann, Kruse, and Rottinger 2000; Kuper and Kroepelin 2006). Thus, migration through the Sahara, and interaction among populations living in different parts of the Sahara, would have been possible approximately every 20 ka, with the Western and Mediterranean margins of North Africa becoming more favorable habitats than the rest of the region during the relatively minor humid phases following MIS 5 and prior to the Holocene.

CONCLUSIONS

Climate variability during Pleistocene time in North Africa would certainly have been a significant stress on hominins living in what is currently an extremely arid environment. Though desiccation events occurred cyclically during at least the last several hundred thousand years, the events varied significantly in the magnitude of climate change, and in the extent to which the changes affected the entire region. During the generally humid MIS 5, drier conditions seem to have affected portions of North Africa; these climate crises could have been dealt with simply by intraregional migration. Regional desiccation at the end of MIS 5, however, would have necessitated population movements either to refugia (coastal region, oases, etc.) or out of North Africa entirely. Throughout most of North Africa, later Pleistocene humid events would have allowed for re-expansion of populations, but water resources would have been more restricted than during MIS 5, perhaps represented only by local, spring-supported wetlands or playas rather than extensive lakes. If populations were smaller during these times than during MIS 5, it may have been possible for refugia to accommodate a large fraction of the population that had formerly inhabited a greener desert. Hominin populations may also have exploited the relatively habitable western Sahara during this time. Additional archaeological survey in areas away from current oases is needed to understand patterns of occupation throughout the whole of the Sahara.

REFERENCES

Adelsberger, K. A., and J. R. Smith. In review. Sedimentology, geomorphology, and paleoenvironmental interpretation of spring-deposited ironstones and associated sediments, Dakhleh Oasis, Western Desert, Egypt. *Catena.*

Akdim, B., and R. Julia. 2005. The travertine mounds of Tafilalet (Morocco): morphology and genesis based on present-day analogues. *Zeitschrift für Geomorphologie* 49: 373–89.

——. 2006. Commentaire a la note intitulee Datation par la methode U/Th d'un travertin quaternaire du Sud-Est Marocain: implications paleoclimatiques pendant le Pleistocene moyen et superieur de L. Boudad et al. [C. R. Geoscience 335 (2003): 469–78]. *Comptes Rendus Geoscience* 338: 581–82.

Armitage, S. J., N. A. Drake, S. Stokes, A. El-Hawat, M. J. Salem, K. White, P. Turner, and S. J. McLaren. 2007. Multiple phases of North African humidity recorded in lacustrine sediments from the Fazzan Basin, Libyan Sahara. *Quaternary Geochronology* 2: 181–86.

Arz, H. W., F. Lamy, J. Patzold, P. J. Muller, and M. Prins. 2003. Mediterranean Moisture Source for an Early-Holocene Humid Period in the Northern Red Sea. *Science* 300: 118–21.

Barton, R. N. E., A. Bouzouggar, S. N. Collcutt, J. L. Schwenninger, and L. Clark-Balzan. 2009. OSL dating of the Aterian levels at Dar es-Soltan I (Rabat, Morocco) and implications for the dispersal of modern Homo sapiens. *Quaternary Science Reviews* 28: 1914–31.

Boudad, L., L. Kabiri, S. Farkh, C. Falgueres, L. Rousseau, J. Beauchamp, E. Nicot, and G. Cairanne. 2003. Datation par la methode U/Th d'un travertin quaternaire du Sud-Est Marocain: implications paléoclimatiques pendant le Pleistocène moyen et supérieur. *Comptes Rendus Geoscience* 335: 469–78.

———. 2006. Reponse au commentaire de Brahim Akdim et Ramon Julia sur la note Datation par la méthode U/Th d'un travertin quaternaire du Sud-Est marocain : implications paléoclimatiques pendant le Pleistocène moyen et supérieur [*C. R. Géoscience* 335 (2003): 469–78]. *Comptes Rendus Geoscience* 338: 583.

Brook, G. A., N. S. Embabi, M. M. Ashour, R. L. Edwards, H. Cheng, J. B. Cowart, and A. A. Dabous. 2003. Quaternary environmental change in the Western Desert of Egypt: evidence from cave speleothems, spring tufas, and playa sediments. *Zeitschrift für Geomorphologie* 131: 59–87.

Brookes, I. A. 1993. Geomorphology and Quaternary geology of the Dakhla Oasis region, Egypt. *Quaternary Science Reviews* 12: 529–52.

Caton-Thompson, G. 1952. *Kharga oasis in prehistory*. London: Athlone Press.

Causse, C., R. Coque, J. C. Fontes, F. Gasse, E. Gibert, H. Ben Ouezdou, and K. Zouari. 1989. Two high levels of continental waters in the southern Tunisian chotts at about 90 and 150 ka. *Geology* 17: 922–25.

Causse, C., B. Ghaleb, N. Chkir, K. Zouari, H. Ben Ouezdou, and A. Mamou. 2003. Humidity changes in southern Tunisia during the late Pleistocene inferred from U-Th dating of mollusc shells. *Applied Geochemistry* 18: 1691–1703.

Churcher, C. S., and M. R. Kleindienst. In press. Great Lakes in the Dakhleh Oasis: mid-Pleistocene freshwater lakes in the Dakhleh Oasis depressions, Western Desert Egypt. In *The Oasis Papers IV: Proceedings of the Fourth International Conference of the Dakhleh Oasis Project*, ed. A. J. Mills. Oxford: Oxbow Books.

Churcher, C. S., M. R. Kleindienst, and H. P. Schwarcz. 1999. Faunal remains from a middle Pleistocene lacustrine marl in Dakhleh Oasis, Egypt: palaeoenvironmental reconstructions. *Palaeogeography, Palaeoclimatology, Palaeoecology* 154: 301–12.

Crombie, M. K., R. E. Arvidson, N. C. Sturchio, Z. El Alfy, and K. Abu Zeid. 1997. Age and isotopic constraints on Pleistocene pluvial episodes in the Western Desert, Egypt. *Palaeogeography, Palaeoclimatology, Palaeoecology* 130: 337–55.

Drake, N. A., A. S. El-Hawat, P. Turner, S. J. Armitage, M. J. Salem, K. H. White, and S. McLaren. 2008. Palaeohydrology of the Fazzan Basin and surrounding regions: the last 7 million years. *Palaeogeography, Palaeoclimatology, Palaeoecology* 263: 131–45.

Geb, M. 2000. Factors favouring precipitation in North Africa: seen from the viewpoint of present-day climatology. *Global and Planetary Change* 26: 85–96.

Geyh, M. A., and F. Thiedig. 2008. The Middle Pleistocene Al Mahruqah Formation in the Murzuq Basin, northern Sahara, Libya: evidence for orbitally-forced humid episodes during the last 500,000 years. *Palaeogeography, Palaeoclimatology, Palaeoecology* 257: 1–21.

Giegengack, R. 1968. *Late Pleistocene history of the Nile Valley in Egypt and Nubia.* Ph.D. diss., Yale University.

Hamdan, M. Abd Elrahman. 2000. Quaternary travertines of wadis Abu Had-Dib area, Eastern Desert, Egypt: paleoenvironment through field, sedimentology, age, and isotopic study. *Sedimentology of Egypt* 8A: 49–62.

Hawkins, A. L. 2001. *Getting a handle on tangs: a study in surface archaeology from the Western Desert of Egypt.* Ph.D. diss., University of Toronto.

Hawkins, A. L., J. R. Smith, R Giegengack, M. A. McDonald, M. R. Kleindienst, H. P. Schwarcz, C. S. Churcher, M. F. Wiseman, and K. Nicoll. 2001. New research on the prehistory of the escarpment in Kharga Oasis. *Nyame Akuma* 55: 8–14.

Haynes, C. V., T. A. Maxwell, A. El Hawary, K. A. Nicoll, and S. Stokes. 1997. An Acheulian site near Bir Kiseiba in the Darb el Arba'in Desert, Egypt. *Geoarchaeology* 12: 819–32.

Hemleben, C., D. Meischner, R. Zahn, A. Almogi, H. Erlenkeuser, and B. Hiller. 1996. Three hundred eighty thousand year long stable isotope and faunal records from the Red Sea: influence of global sea level change on hydrography. *Paleoceanography* 11: 147–56.

Hill, C. L. 2001. Geologic contexts of the Acheulian (Middle Pleistocene) in the Eastern Sahara. *Geoarchaeology* 16: 65–94.

Hoelzmann, P., H.-J. Kruse, and F. Rottinger. 2000. Precipitation estimates for the eastern Saharan palaeomonsoon based on a water balance model of the West Nubian palaeolake basin. *Global and Planetary Change* 64: 105–20.

Issar, A. 2003. *Climate changes during the Holocene and their impact on hydrological systems, International Hydrology Series.* Cambridge: Cambridge University Press.

Kieniewicz, J. M., and J. R. Smith. 2009. Paleoenvironmental reconstruction and water balance of a Mid-Pleistocene pluvial lake, Dakhleh Oasis, Egypt. *Geological Society of America Bulletin* 121: 1154–71.

———. 2007. Hydrologic and climatic implications of stable isotope and minor element analyses of authigenic calcite silts and gastropod shells from a mid-Pleistocene pluvial lake, Western Desert, Egypt. *Quaternary Research* 68: 431–44.

Kleindienst, M. R. 1999. Pleistocene archaeology and geoarchaeology of the Dakhleh Oasis: a status report. In *Reports from the survey of the Dakhleh Oasis 1977–1987*, ed. C. S. Churcher and A. J. Mills. Oxford: Oxbow Books.

Kleindienst, M. R., H. P. Schwarcz, K. Nicoll, C. S. Churcher, J. Frizano, R. Giegengack, and M. F. Wiseman. 2008. Water in the desert: first report on uranium-series dating of Caton-Thompson's and Gardner's "classic" Pleistocene sequence at Refuf Pass, Kharga Oasis. In *Oasis Papers II: Proceedings of the Second International Conference of the Dakhleh Oasis Project*, ed. M. F. Wiseman. Oxford: Oxbow Books.

Kuper, R., and S. Kroepelin. 2006. Climate-controlled Holocene occupation in the Sahara: motor of Africa's evolution. *Science* 313: 803–7.

Larrasoaña, J. C., A. P. Roberts, E. J. Rohling, M. Winkhofer, and R. Wehausen. 2003. Three million years of monsoon variability over the northern Sahara. *Climate Dynamics* 21: 689–98.

Legge, H.-L., J. Mutterlose, H. W. Arz, and J. Pätzold. 2008. Nannoplankton successions in the northern Red Sea during the last glaciation (60 to 14.5 ka BP): reactions to climate change. *Earth and Planetary Science Letters* 270: 271–79.

McBurney, C. B. M. 1967. *The Haua Fteah (Cyrenaica) and the stone age culture of the south-east Mediterranean.* Cambridge: Cambridge University Press.

McCauley, J. F., C. S. Breed, G. G. Schaber, W. P. McHugh, B. Issawi, C. V. Haynes Jr., C. Elachi, and R. Blom. 1982. Subsurface valleys and geoarchaeology of the Eastern Sahara revealed by Shuttle radar. *Science* 218: 1004–20.

McHugh, W. P., C. S. Breed, G. G. Schaber, J. F. McCauley, and B. J. Szabo. 1988. Acheulian sites along the "radar rivers," southern Egyptian Sahara. *Journal of Field Archaeology* 15: 361–79.

McHugh, W. P., J. F. McCauley, C. V. Haynes Jr., C. S. Breed, and G .G. Schaber. 1988. Paleorivers and geoarchaeology in the southern Egyptian Sahara. *Geoarchaeology* 3: 1–40.

McKenzie, J. A. 1993. Pluvial conditions in the eastern Sahara following the penultimate deglaciation: implications for changes in atmospheric circulation patterns with global warming. *Palaeogeography, Palaeoclimatology, Palaeoecology* 103: 95–105.

Mercier, N., L. Wengler, H. Valladas, J. L. Joron, L. Froget, and J. L. Reyss. 2007. The Rhafas Cave (Morocco): chronology of the mousterian and aterian archaeological occupations and their implications for Quaternary geochronology based on luminescence (TL/OSL) age determinations. *Quaternary Geochronology* 2: 309–13.

Moreno, A., J. Targarona, J. Henderiks, M. Canals, T. Freudenthal, and H. Meggers. 2001. Orbital forcing of dust supply to the North Canary Basin over the last 250 kyr. *Quaternary Science Reviews* 20: 1327–39.

Nespoulet, R., M. A. El Hajraoui, F. Amani, A. Ben Ncer, A. Debénath, A. El Idrissi, J.-P. Lacombe, P. Michel, A. Oujaa, and E. Stoetzel. 2008. Palaeolithic and Neolithic occupations in the Témara region (Rabat, Morocco): recent data on hominin contexts and behavior. *African Archaeological Review* 25: 21–39.

Nicoll, K., R. Giegengack, and M. Kleindienst. 1999. Petrogenesis of artifact-bearing fossil-spring tufa deposits from Kharga Oasis, Egypt. *Geoarchaeology* 14: 849–63.

Olszewski, D. I., H. L. Dibble, U. A. Schurmans, S. P. McPherron, and J. R. Smith. 2005. High desert Paleolithic survey at Abydos (Egypt): preliminary report. *Journal of Field Archaeology* 30: 283–303.

Osborne, A. H., D. Vance, E. J. Rohling, N. Barton, M. Rogerson, and N. Fello. 2008. A humid corridor across the Sahara for the migration of early modern humans out of Africa 120,000 years ago. *Proceedings of the National Academy of Sciences* 105: 16444–47.

Osmond, J. K., and A. A. Dabous. 2004. Timing and intensity of groundwater movement during Egyptian Sahara pluvial periods by U-series analysis of secondary U in ores and carbonates. *Quaternary Research* 61: 85–94.

Paillou, P., M. Schuster, S. Tooth, T. Farr, A. Rosenqvist, S. Lopez, and J.-M. Malezieux. 2009. Mapping of a major paleodrainage system in eastern

Libya using orbital imaging radar: the Kufrah River. *Earth and Planetary Science Letters* 277: 327–33.

Rohling, E. J., T. R. Cane, S. Cooke, M. Sprovieri, I. Bouloubassi, K. C. Emeis, R. Schiebel, D. Kroon, F. J. Jorissen, A. Lorre, and A. E. S. Kemp. 2002. African monsoon variability during the previous interglacial maximum. *Earth and Planetary Science Letters* 202 (1): 61–75.

Schwarcz, H. P., and L. Morawska. 1998. Uranium-series dating of carbonates from Bir Tarfawi and Bir Sahara East. In *Egypt during the last Interglacial*, ed. F. Wendorf, R. Schild, and A. E. Close. New York: Plenum Press.

Smith, J. R. 2010. Palaeoenvironments of eastern North Africa and the Levant in the late Pleistocene. In *South-Eastern mediterranean peoples between 130,000 and 10,000 years ago*, ed. E. A. A. Garcea. Oxford: Oxbow Books.

———. In press. Spatial and temporal variation in the nature of Pleistocene pluvial phase environments across North Africa. In *Modern human origins: a North African perspective*, ed. J.-J. Hublin and S. McPherron. New York: Springer-Verlag.

Smith, J. R., and R. Giegengack. 2003. Spatial and temporal distribution of fossil-spring tufa deposits, Western Desert, Egypt: Implications for palaeoclimatic interpretations. In *Oasis Papers III: Proceedings of the Third International Conference of the Dakhleh Oasis Project*, ed. G. E. Bowen and C. A. Hope. Oxford: Oxbow Books.

Smith, J. R., R. Giegengack, and H. P. Schwarcz. 2004. Constraints on Pleistocene pluvial climates through stable-isotope analysis of fossil-spring tufas and associated gastropods, Kharga Oasis, Egypt. *Palaeogeography, Palaeoclimatology, Palaeoecology* 206: 157–75.

Smith, J. R., R. Giegengack, H. P. Schwarcz, M. M. A. McDonald, M. R. Kleindienst, A. L. Hawkins, and C. S. Churcher. 2004. A reconstruction of Quaternary pluvial environments and human occupations using stratigraphy and geochronology of fossil-spring tufas, Kharga Oasis, Egypt. *Geoarchaeology* 19: 407–39.

Smith, J. R., A. L. Hawkins, Y. Asmerom, V. Polyak, and R. Giegengack. 2007. New age constraints on the Middle Stone Age occupations of Kharga Oasis, Western Desert, Egypt. *Journal of Human Evolution* 52: 690–701.

Smith, J. R., J. Kieniewicz, and Y. Asmerom. submitted. Terrestrial record of Saharan humidity and habitability during the Pleistocene. *Geology*.

Stein, M., A. Almogi-Labin, S. L. Goldstein, C. Hemleben, and A. Starinsky. 2007. Late Quaternary changes in desert dust inputs to the Red Sea and Gulf of Aden from $^{87}Sr/^{86}Sr$ ratios in deep-sea cores. *Earth and Planetary Science Letters* 261: 10419.

Sturchio, N. C., X. Du, R. Purtschert, B. E. Lehmann, M. Sultan, L. J. Patterson, Z. T. Lu, P. Mueller, T. Bigler, K. Bailey, T. P. O'Connor, L. Young, R. Lorenzo, R. Becker, Z. El Alfy, B. El Kaliouby, Y. Dawood, and A. M. A. Abdallah. 2004. One million year old groundwater in the Sahara revealed by krypton-81 and chlorine-36. *Geophysical Research Letters* 31: 1–4.

Sultan, M., N. C. Sturchio, F. A. Hassan, M. A. R. Hamdan, A. M. Mahmood, Z. El Alfy, and T. Stein. 1997. Precipitation source inferred from stable

isotopic composition of Pleistocene groundwater and carbonate deposits in the Western Desert of Egypt. *Quaternary Research* 48: 29–37.

Szabo, B. J., C. V. Haynes Jr., and T. A. Maxwell. 1995. Ages of Quaternary pluvial episodes determined by uranium-series and radiocarbon dating of lacustrine deposits of Eastern Sahara. *Palaeogeography, Palaeoclimatology, Palaeoecology* 113: 227–41.

Thorweihe, U. 1990. Nubian aquifer system. In *The geology of Egypt*, ed. R. Said. Rotterdam: A. A. Balkema.

Van Peer, P. 1998. The Nile corridor and the Out-of-Africa model: an examination of the archaeological record. *Current Anthropology* 39: S115–S140.

Van Peer, P., R. Fullagar, S. Stokes, R. M. Bailey, J. Moeyersons, F. Steenhoudt, A. Geerts, T. Vanderbeken, M. De Dapper, and F. Geus. 2003. The early to middle Stone Age transition and the emergence of modern human behaviour at site 8-B-11, Sai Island, Sudan. *Journal of Human Evolution* 45: 187–93.

Vermeersch, P. M., ed. 2000. *Palaeolithic living sites in upper and middle Egypt.* Ithaca: Cornell University Press.

Vermeersch, P. M., M. Otte, E. Gilot, E. Paulissen, G. Guselings, and D. Drappier. 1982. Blade technology in the Egyptian-Nile Valley: some new evidence. *Science* 216: 626–28.

Warner, T. T. 2004. *Desert meteorology.* Cambridge: Cambridge University Press.

Weisrock, A., L. Wengler, J. Mathieu, A. Ouammou, M. Fontugne, N. Mercier, J. L. Reyss, H. Valladas, and P. Guery. 2006. Upper Pleistocene comparative OSL, U/Th and (super 14) C datings of sedimentary sequences and correlative morphodynamical implications in the south-western Anti-Atlas (Oued Noun, 29 degrees N, Morocco). *Quaternaire* 17: 45–59.

Weldeab, S., D. W. Lea, R. R. Schneider, and N. Andersen. 2007. 155,000 years of West African monsoon and ocean thermal evolution. *Science* 316 (5829): 1303–7.

Wendorf, F., and R. Schild, eds. 1976. *The prehistory of the Nile Valley.* New York: Academic Press.

Wendorf, F., R. Schild, and A. E. Close, eds. 1993. *Egypt during the last Interglacial: the middle Paleolithic of Bir Tarfawi and Bir Sahara East.* New York: Plenum Press.

Wengler, L., A. Weisrock, J.-E. Brochier, J.-P. Brugal, M. Fontugne, F. Magnin, J. Mathieu, N. Mercier, A. Ouammou, J.-L. Reyss, F. Senegas, H. Valladas, and L. Wahl. 2002. Enregistrement fluviatile et paleoenvironnements au Pleistocene superieur sur la bordure atlantique de l'Anti-Atlas (Oued Assaka, S-O marocain). *Quaternaire* 13 (3-4): 179–92.

Zouari, K., N. Chkir, and C. Causse. 1998. Pleistocene humid episodes in southern Tunisian chotts. Isotope techniques in the study of past and current environmental changes in the hydrosphere and the atmosphere: *Proceedings of an international symposium on applications of isotope techniques in studying past and current environmental changes in the hydrosphere and the atmosphere.* International Atomic Energy Agency (Vienna): 543–54.

8

Late Middle Holocene Climate and Northern Mesopotamia: Varying Cultural Responses to the 5.2 and 4.2 ka Aridification Events

MICHAEL D. DANTI

INTRODUCTION

The cultural responses to proposed Middle Holocene climate changes in the Near East have been much studied and debated the last twenty-five years (Bottema 1989; Sanlaville 1992; Roberts and Wright 1993; Courty 1994; Butzer 1995; Brooks 2006; Rosen 2007), especially in relation to cultural complexity in northern Mesopotamia (Hole 1994; Weiss 2000; Zettler 2003; Issar and Zohar 2004; Staubwasser and Weiss 2006; Cooper 2006; Schwartz and Miller 2007; Kuzucuğlu and Marro 2007). A marked increase in proxy datasets of global and regional climate and the availability of more holistic and detailed archaeological coverage have driven this trend. As often is the case with scientific advances, scholars have cited climate change as the prime suspect for virtually every major cultural development in the Late Quaternary of Greater Mesopotamia, including plant domestication (Moore and Hillman 1992), the development of complex society (Hole 1994), and the genesis and termination of nearly every major cultural horizon.[1]

Occasionally, the overzealous application of paleoclimatic datasets by archaeologists, or by paleoclimatologists armed only with general syntheses on prehistory to map culture changes, evokes the famous quote by Mark Twain regarding science, "One gets such wholesale returns of conjecture out of such a trifling investment of fact" (Twain 1883). Nevertheless, the availability of increasingly more detailed paleoclimatic data and, in particular, the revelation that Holocene climate was not as stable as once assumed, have had a major impact on our understanding of northern Mesopotamia. Still, much work needs to be done to understand the cultural responses to climatic stresses and their ultimate effects, especially for the Middle Holocene, since it was during this time that cultural complexity emerged in the Near East. In particular, we must focus more on the structure of subsistence economies in the *longue durée*

*I would like to thank Naomi Miller and Richard Zettler for reading drafts of this paper and providing many useful suggestions.

and regions that seem to defy the trend of "collapse," for it is in these areas that important lessons were learned regarding how to cope with the changing environment. These adaptations would profoundly influence subsequent periods. Moreover, we must stop trying to force universal models of cultural responses onto all regions of northern Mesopotamia when no universal pattern is emerging with continued, and increasingly more focused, research. The Middle Euphrates region of northern Syria, and in particular the Early Bronze Age (EBA) site of Tell es-Sweyhat, is one such area where generalized interpretations of late Middle Holocene collapses caused by climate events do not fit. I contend that in this area, and much of western Syria, an emphasis on pastoral production and the smaller scale of settlement systems provided a more resilient socioeconomic base for sustaining urbanism during the climate fluctuations of the later Middle Holocene. The sudden and heightened impact of the 4.2 ka aridification event beginning ca. 2200 BC on the Khabur region in northeastern Syria, while well documented, is anomalous, and was perhaps due to an imbalanced production strategy of intensive dry farming related to its incorporation into the Akkadian Empire, which created a brittle socioeconomic system, weakened by policies mandating surplus production for a foreign power. Hints of adaptive responses are readily apparent in the late third millennium BC, where we see a steadily increasing shift toward transhumant pastoralism and concomitant decline in urbanism in northern Mesopotamia, while in southern Mesopotamia the Ur III "empire" or territorial state exhibited a previously unattested obsessive preoccupation with controlling pastoral production in its hinterland.

Incorporating Climate into Archaeological Interpretations of Northern Mesopotamia

Incorporating climate data into interpretations of proto-historical and early historical periods is a particularly daunting proposition, and one finds that historically attested climate catastrophes have a chimeric quality, especially for the late Middle Holocene and early Late Holocene of northern Mesopotamia. The task is hardly easier with an increase in historical documentation in later periods, and for those seeking ethnohistorical analogies there is a disturbing trend that the impact of climate events seems inversely proportionate to the detail provided by historical documentation. While droughts and famines are attested in Middle Eastern historical sources, one is hard pressed to find contemporary accounts connecting them to mass migrations, urban collapse, and the downfall of political regimes. Despite these challenges, in recent years scholars have been grappling with catastrophic climate events in their interpretations of the rise of cultural complexity in the Near East, in particular the events of 6.2 ka, 5.2 ka, and 4.2 ka. The latter two events, characterized by aridity, bracket the EBA of the Near East (3100–2000 BC). The 5.2 ka event is often interpreted as contributing to the abrupt collapse of the Uruk (4000–3100 BC), the world's first known civilization (Brooks 2006; Weiss 2003; Staubwasser and Weiss 2006). The 4.2 ka event is believed to have

brought down the world's first experiment with empire, the Akkadian Empire (2350–2150 BC), and instigated a widespread decline in urbanism throughout northern Mesopotamia (Weiss et al. 1993; Cullen et al. 2000; deMenocal 2001; Staubwasser et al. 2003; Drysdale et al. 2006; Arz, Lamy, and Pätzold 2006; Staubwasser and Weiss 2006). The intervening period was marked by an amelioration of aridity.

Two theoretical stances, often verging on mutually exclusive positions in scholarly debate, have emerged to explain the de-urbanization of northern Mesopotamia at the end of the EBA, and by extension other Holocene climate events. The "Catastrophic Collapse Model" stresses the abrupt onset, magnitude, and long duration of the climate events (Weiss et al. 1993; Staubwasser and Weiss 2006). The "Brittle Economy Model" posits that the maximization of agricultural outputs, driven by population growth and attendant breakdowns in risk-management strategies, led to the development of vulnerable agricultural systems incapable of absorbing and adjusting to climate-induced stresses, resulting in prolonged periods of decline and collapse (Wilkinson 1994, 1997, 2004).[2] The Brittle Economy Model is in part a corollary of theories that maintain that documented anthropogenic environmental degradation caused by urbanization and agriculture might have lowered the resilience of urban centers to exogenous stresses (Adams 1978; Miller 1990, 1997a; McCorriston 1995). The important difference between the Catastrophic Collapse Model and the Brittle Economy Model is the scale of the climate event capable of destabilizing states. The Brittle Economy Model contains inherent, growing instabilities driven by population pressure, abandonment of risk-management strategies, and environmental degradation. Catastrophic events can, of course, induce collapse, but so can climatic fluctuations in the "normal" range of climatically stable time periods (a series of bad years within the normal range of interannual variation in precipitation). The two models also differ in terms of the archaeological datasets on which they were ultimately based. The Catastrophic Collapse Model of the 4.2 ka event primarily derives from archaeological work in the Khabur region, while the Brittle Economy Model was originally developed from regional research in the area of the Middle Euphrates. Proxy data on climate is scarce in both these regions, and reconstructions of climate are largely based on work done in surrounding areas. In terms of reliability, scholars have generally settled on the Soreq Cave speleothems and Lake Van varve sequence as the most reliable climate proxies (Bar-Matthews and Ayalon 1997; Bar-Matthews, Ayalon, and Kaufman 1998; Bar-Matthews et al. 1999; Kaufman et al. 2003; Bar-Matthews and Ayalon 2004; Lemcke and Sturm 1997). In part, regional differences in the structure of agropastoral economies, particularly localized constraints/capabilities, have influenced the development of these theoretical stances. While annual precipitation is quite similar across latitudinal zones in the Middle Euphrates and Khabur regions, the Khabur Plains are more suitable for agricultural intensification. The region provides much more arable land for agriculture in comparison to the small embayments of the Middle Euphrates region, which are circumscribed by high limestone plateaus and undulating steppe that were largely unsuitable for dry farming prior to the

introduction of modern farming techniques. Consequently, the EBA urban centers of the Middle Euphrates region never attained the sizes of those of the Khabur Plains (Wilkinson 1994) and emphasized sheep/goat pastoralism as a means for utilizing the surrounding unarable land as seasonal pasture. The Khabur Plain also differed from the Middle Euphrates region in that the former exhibited much closer connections to southern Mesopotamia throughout history. Control of the Khabur afforded its southern neighbors dramatically increased agricultural productivity and important transportation routes.

While "collapse" and "decentralization" aptly summarize the aftermath of the major climate events of the Middle Holocene for some subregions of northern Mesopotamia and western Syria, such terms have virtually precluded the study of adaptive responses at lower scales of analysis. Many state systems survived the initial impact of the 4.2 ka event with only minor perturbations or even flourished for another 100–200 years.[3] Therefore, certain responses or existing factors supported some groups during these crises in northern Mesopotamia and would have profoundly shaped developments in subsequent periods. Conversely, some factor(s) might have increased the vulnerability of certain states to environmental stresses. The neglect of these research topics stems from some scholars' desire to emphasize the catastrophic, almost inexorable, and universal impact of the 4.2 ka event, at times to the extent that contradictory data are relegated to the status of irrelevant anomaly. This obviates the need to posit any potential cultural responses or inherent adaptive mechanisms beyond wholesale "collapse": recovery occurs when climate improves. For northern Mesopotamia, my own impression is that some scholars view the potential of certain state systems to survive proposed catastrophic climate events as a direct challenge to the scientific evidence for the events and their severity—archaeological data are transformed into proxy climate data. Researchers who work at sites that withstand catastrophes can at times find themselves in an odd defensive position for want of archaeological evidence for disruptions, defending evidence for late EBA prosperity. In the Catastrophic Model, only cursory references are typically made to the aftereffects of climate-induced collapse; in the case of northern Mesopotamia this usually involved subsequent large-scale migrations to the irrigated alluvial plain of southern Mesopotamia, precipitating a domino effect of political disintegration. In the same vein, the Brittle Economy Model's emphasis on the precarious condition of some EBA state systems often begs the question of their genesis and long-term success in seeming defiance of climatic phenomena—both short-term fluctuations and sudden punctuated events.

The largest fly in the ointment for both models is the EBA site of Tell es-Sweyhat, located in northern Syria, and similar sites in the Middle Euphrates region (Fig. 1). The Sweyhat region seems to have thrived when climate deteriorated in the Middle Holocene.[4] In the late fourth to early third millennia, as the Uruk faded, a new and seemingly unlikely economic strategy took hold, focused on agropastoral production in agriculturally marginal areas. This system of production balanced the dry farming of barley with sheep/

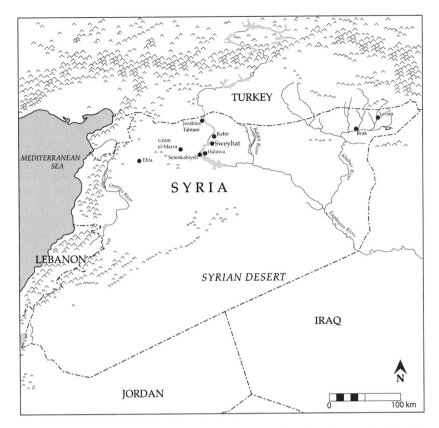

Figure 1. Map of Syria showing the location of Tell es-Sweyhat and other sites mentioned in the text.

goat pastoralism based on a low degree of transhumance, and much of the barley production was driven by the need for conserved fodder (barley and straw) for herds. Balanced agropastoralism lies amid a spectrum of production strategies historically practised in the region defined by dependence on conserved fodder, which is inversely proportionate to the degree of pastoral transhumance. Alternatives include dimorphic agropastoralism, in which herding and agriculture are more specialized activities, primarily driven by higher degrees of transhumance with concomitant degrees of sociopolitical separation between agricultural and pastoral producers, and on the far end of the spectrum true nomadism. Again, when the Akkadian Empire disintegrated in the later millennium and agrarian states floundered elsewhere, Tell es-Sweyhat and many of its neighbors flourished. How could Sweyhat and similar urban centers have had the very definition of brittle economies and at the same time seemingly prospered during the catastrophic climate events of the Middle Holocene while more agriculturally productive areas experienced widespread urban collapse?

THE TELL ES-SWEYHAT REGION

Tell es-Sweyhat is located in the Middle Euphrates region of northern Syria (Fig. 2).[5] The site sits on the left bank of the Euphrates at the center of a crescent-shaped embayment, a mid- to late Pleistocene terrace carved out of the high plateau by an arm of the river (deHeinzelin 1967; Wilkinson 2004, 19–20). Historically, such embayments have been important foci of settlement (Besançon and Sanlaville 1985, 13). The three major environmental zones are the floodplain, the main terrace, and the high plateau. The settlement was occupied continuously from at least the early third until the early second millennium BC, and consists of a high, central mound (H. 14.5 m) surrounded by an extensive low mound of 35–45 ha. Following the site's abandonment in the early second millennium it was not again occupied until the Seleucid and Late Roman periods (second century BC–AD third century), when a small village was established on the earlier mound. The attested periods of occupation at Tell es-Sweyhat represent peak periods of settlement in the Sweyhat embayment (Wilkinson 2004) and to the east on the high plateau (Danti 2000).

Tell es-Sweyhat and a few other neighboring ancient cities were conspicuous and unique in that they were situated far from the river, 5 km in the case of Sweyhat, in a marginal agricultural environment on the unirrigable main terrace. Today the region averages 250 mm of precipitation per annum with an interannual variability of 25–35% (Wallén 1968, 226–31). This is just enough rainfall to make the dry farming of drought-resistant varieties of barley viable with two successful crops every three years (Al-Ashram 1990, 167). This situation places the Sweyhat region near the southern limit for rainfall agriculture today (Perrin de Brichambaut and Wallén 1963).

THE 5.2 KA EVENT AND MIDDLE HOLOCENE PLUVIAL IN THE SWEYHAT REGION

The history of Middle Holocene settlement in the Sweyhat region is well documented and often runs counter to initial expectations based on paleoclimatic data. In the broader Middle Euphrates region, the Late Chalcolithic (LC) period is characterized by a phase of urbanization and increased cultural complexity spurred by close connections with, and colonization by, southern Mesopotamian Uruk populations, which ended abruptly at the end of the fourth millennium BC with political decentralization and a return to regionalized cultural traditions (Akkermans and Schwartz 2003, 209–11). As previously mentioned, scholars have recently attributed this collapse to aridity brought on by the 5.2 ka climate event (Weiss and Bradley 2001), which caused "the abrupt Late Uruk colony collapse across Greater Mesopotamia, extending from the Zagros to Syria, as precipitation in rainfed cereal agriculture regions was reduced beyond sustainable limits" (Staubwasser and Weiss 2006, 379). Concurrent with this event, i.e., the terminal late Uruk period and LC-EBA transition (Hajji Ibrahim Period A in the Sweyhat area, cf. Table 1), the main terrace in the Sweyhat region was colonized (Danti 2000; Wilkinson

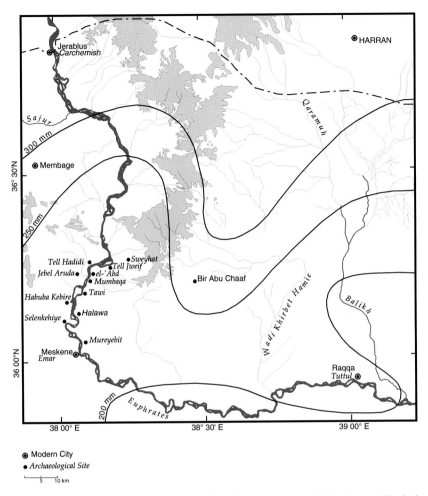

Figure 2. Map of Tell es-Sweyhat region showing annual rainfall isohyets. Shaded areas represent elevations over 500 m (AMSL).

2004, 179).[6] This new, albeit extremely modest, settlement system (sites on the order of 0.25 ha) lay relatively far from the Euphrates (more than 5 km) and was directed at agropastoral production. The excavation of Tell Hajji Ibrahim (Danti 2000),[7] located ca. 1 km southeast of Tell es-Sweyhat, uncovered a small, fortified compound with one house and granaries, providing storage far in excess of the annual needs of a single family. The morphology of other small, contemporary mounds indicates that they likely represent the remains of similar facilities. Plant remains from Hajji Ibrahim strongly suggest the cleaning and storage of barley although the two samples published to date come from the area of a Period B bread oven (see below), that is, the period when the granaries fell out of use (Miller 1997b, 103–4). Thick deposits of charred

plant remains were recovered from the final two Period A2 floors of the Main House and contained high numbers of barley internodes, probably indicative of crop processing debris (Danti 2000, 115). Excavations at Tell es-Sweyhat, the major site in the region in later time periods, have not located occupation levels contemporary with Hajji Ibrahim Period A.[8] Tell Hajji Ibrahim and similar sites were almost certainly connected to a larger local population of transhumant agropastoralists (Danti 1997, 2000; Danti and Zettler 1998). In staple-finance models, such granaries might be interpreted as surplus-producing satellite settlements in a redistributive network centered on a large urban center (D'Altroy and Earle 1985). This is highly unlikely in the case of the Sweyhat region in Hajji Ibrahim Period A since no urban centers existed in this region at this time. The long-distance overland transport of grain by donkeys was prohibitively expensive and seems to have been seldom practiced since the animals required large amounts of high protein fodder (grain). Riverine transport was practiced in Mesopotamia, but the treacherous upper courses of the Middle Euphrates made the transport of bulk commodities risky.

The simple answer regarding the purpose of the Tell Hajji Ibrahim storage facilities relates them to the annual pastoral cycle. Large-scale sheep and goat production in northern Mesopotamia requires conserved fodder in the form of grain and straw to feed animals in the cold and wet winter and spring, when there is virtually no pasture of nutritive value and when the caloric and, especially, protein requirements of small ruminants are elevated due to the inclement weather and especially the requirements of adult females that reach late gestation and lactation at this time during the birthing and milking seasons (Rihawi et al. 1987, 452; Thomson and Bahhady 1995, 7).[9] The critical importance of barley straw and grain as a conserved fodder is best illustrated by the evolution of the Syrian landrace *Arabic abiad*, which is drought resistant and, most tellingly, produces a straw that is relatively high in protein compared to other barley varieties (Capper et al. 1985a, 1985b; Cocks 1986; Nygaard 1983). All too often, grain production is seen purely in terms of human diet, but barley grain and straw, especially in the marginal zone of northern Mesopotamia, were requirements for sustainable pastoral production. The human population of northern Mesopotamia almost certainly preferred wheat. Another widely held misconception concerns the range of transhumance in the annual cycle of pastoralists in antiquity. Before the development of nomadism in the later second millennium BC, concurrent with the introduction of the domesticated camel, pastoralists in northern Mesopotamia, like the so-called sheep-tribes of the modern era, almost certainly practised short-distance transhumance unless severe stress forced them out of their tribal territories (d'Hont 1994, 209–11; Oppenheim 1939, 208–17; Charles 1939). During the cold season, there simply is nothing of nutritional value in the scant pasture of northern Mesopotamia, and long-distance searches for feed would further stress animals, especially in late winter and early spring during the late gestation stage of females and the lambing season.

Pastoralism based on small ruminants has traditionally been the primary means for offsetting bad harvests and crop failures in the marginal zone of northern Mesopotamia. Syrian landrace sheep, varieties of fat-tailed sheep

like the Awassi (Epstein 1985), are famed for their fat-storing capabilities and were probably the main breed in antiquity (Ryder 1993; Steinkeller 1995). When mobilized, these fat reserves help the animal to survive in lean years (Ghosh and Abichandani 1981, 24–25; Goodchild, El-Awad, and Gürsoy 1999). When crops fail, the unharvested remains serve as in-field fodder rather than simply going to waste and more animals can be slaughtered to offset food deficits and conserve grain for human consumption. Thus, small ruminants minimize risk due to interannual climatic fluctuations and provide valuable contributions to the subsistence economy in the form of meat and dairy products. Perhaps of even more importance was wool production, which provided the local inhabitants with a form of storable wealth for regional and long-distance exchange, as would the production of surplus animals. This economy did not merely enable the inhabitants of the Middle Euphrates region to survive, it would eventually support prosperous city-states.

The succeeding periods, the EB I–III (Sweyhat Periods 1–3, Hajji Ibrahim Periods B–D), are best known from the lowest levels of Tell es-Sweyhat and from the terminal phases of occupation at Tell Hajji Ibrahim. In Hajji Ibrahim Periods B–C (EB I–II) the granaries were demolished, although the site continued to be occupied as a small fortified hamlet. Sometime between ca. 2900 and 2800 BC the site was abandoned, as were most of the other small sites on the main terrace like it (Danti 2000). We preliminarily interpret this as a consolidation of regional settlement and a shift from lineages serving as the units of socioeconomic organization to a system centered on a complex chiefdom. During EBI, Tell es-Sweyhat was a small village, although we have few exposures of these early periods because of the depth of overlying levels. A large fortress was constructed at the center of the site by at least EBII or possibly in EBI, although we have yet to reach its lowest level. The fortress was expanded throughout the EBII and EBIII periods, and eventually measured minimally 75 m east–west by 62 m north–south and at least 3 m high (the corbelled-arch roofs of the internal chambers are still preserved in most places).[10] Architectural remains at the outskirts of the settlement indicate some of the inhabitants practised, minimally, a biseasonal settlement pattern (Armstrong and Zettler 1997, 16–17). The settlement covered a modest 5 ha (Zettler et al. 1997, 169). In EBIII, the fortress was repeatedly enlarged and the settlement slowly grew. Cemeteries of shaft-and-chamber tombs were located on the peripheries of the settlement in the EB III and EB IVa periods.[11] Grave goods attest to a modicum of affluence and access to goods requiring long-distance exchange networks (Zettler et al. 1997, 51–72). During EBIVa (Sweyhat Period 3), the equivalent of the later Early Dynastic III and Old Akkadian period in southern Mesopotamia, the town surrounding the fortress expanded until it covered an estimated 15 ha. This period is not well known at Tell es-Sweyhat since building activities in the following period greatly disturbed Period 3 deposits. In late Period 2 and Period 3, structures of apparent residential nature were built against the sides of the fortress and, in Period 3, eventually on top of it. The narrow chambers inside the massive brickwork of the fortress were filled in to support the structures above, thus fully preserving the building and the corbelled-arched roofs.

In summary, as the Uruk state collapsed we see the beginning of an unprecedented settlement pattern in the Sweyhat region geared toward maximizing returns on agropastoralism—intensive marginal-zone agropastoralism centered on settlements with likely seasonally transhumant elements in the initial stages of the pattern. Grain could be cultivated in the embayment and harvests stored in fortified facilities with small year-round resident populations to guard and maintain them. A similar situation has been advanced for the marginal Middle Khabur region (McCorriston 1995; Hole 1991, 1999). Human diet was likely supplemented by riverine resources such as fish and waterfowl, as in later periods (Weber 1997, 141–42). The riparian floodplain served as a source of wood and a hunting ground or it could be cleared for irrigated summer gardens. Wheat and barley could not be grown here due to the annual flooding of the river in April and May just before grain ripened and could be harvested. Thus, we see a balanced subsistence economy utilizing a number of different resource zones. During this period, there is no archaeological evidence for settlement in the high plateau (Danti 2000). Over time, settlement was centralized at Tell es-Sweyhat (Sweyhat Periods 2–3), which developed into a 10–15 ha fortress town with a fairly affluent population as attested in the graves. To date, there is no evidence from the later phases of the fortress town (Sweyhat Period 3) of serious disruptions associated with a major climate event or Akkadian military campaigns; however, as previously mentioned, deposits from this critical period are fairly rare due to the substantial remodeling of the urban environment at the beginning of Sweyhat Period 4, and only future research can clarify this matter. The most compelling evidence for Tell es-Sweyhat's success in the face of climate change is its growth and development in the late third millennium BC.

THE 4.2 KA EVENT IN THE SWEYHAT REGION

Sweyhat Period 4 begins with a substantial reorganization of the site's urban environment, with the former fortress and town transformed into a fortified Inner City, comprised of two tiers, surrounded by a fortified Outer City.[12] A high terrace stood at the center of the city. The area atop the terrace, the High Inner City, stood one full story above the surrounding Low Inner City. A long-room temple with bent-axis approach and ancillary structures was built atop the High Terrace. The Low Inner City contained various production facilities, storage areas, and residences. The Inner City was protected by a 2.75 m-thick fortification wall with projecting towers. Buildings abutted the interior of this wall and the stone retaining wall of the terrace. Structures of Sweyhat Period 3 were demolished to construct the Inner City fortifications and the High Inner City terrace, and thus far there is no evidence for architectural continuity between Sweyhat Periods 3 and 4 in any area of the site—the change in the urban environment was widespread and radical. On the surrounding plain beyond the Inner City, an outer earthen rampart, fortification walls, and ditch were built to protect the Outer City, enclosing an area measuring roughly 700 by 600 m (Holland 1976, 36; Zettler et al. 1997, 49–51). The Outer City fortifica-

tions covered the entrances to many of the Sweyhat Period 3 tombs. The Outer City consisted of large houses, open spaces, and production facilities (Zettler et al. 1997, 35–51). The settlement continued to grow in the late third millennium BC, eventually expanding beyond the outer fortifications in at least the south, reaching a minimum size of 35–45 ha, but we are unsure of the city's maximum extent, and it should be pointed out that the settlement was not especially dense in the areas of the Outer Town that we have thus far excavated.

While Tell es-Sweyhat dominated the embayment in Sweyhat Period 4, other smaller sites were located on the main terrace and on the fringes of the floodplain (Wilkinson 2004, 138–42). Archaeological survey in the surrounding high plateau has revealed that small settlements were located on pockets of arable land in this challenging environment, usually along major seasonal streambeds (Danti 2000). Modern climate records show that this area actually receives slightly more rainfall than the adjacent area of the Middle Euphrates; however, the water table is quite low, making it difficult to dig wells, and there are no perennial water sources. Traditionally, this area was used as seasonal pasture in spring and early summer, and I maintain that the Sweyhat Period 4 sites on the high plateau, which tend to be less than 5 ha, were likely founded as part of a policy of rangeland improvement—small plots of barley could be raised there for use as fodder, and pastoral stations would have provided shelter and protection for herders and animals attached to the cities lining the Middle Euphrates and the Balikh River to the east.

In late Sweyhat Period 4 or Period 5, at least four buildings in the Inner City were burned: a warehouse (Holland's Area IV), the temple on the High Terrace, and structures in the eastern and southwestern Inner City. It is not clear whether these burnings were simultaneous, but it seems likely. The buildings contained large numbers of *in situ* finds, indicating at least one violent disruption. Following the fire, the buildings were at least partially reconstructed along the same lines. Radiocarbon dates from the burned level of the warehouse include three samples of charcoal, most likely from the roof beams and cleaned grain from a storage jar (Table 1). The charcoal samples date the construction of the building to ca. 2200–2100 BC. The grain tentatively places the destruction of the building toward the end of the third millennium BC. The buildings located along the southwestern and eastern Inner Fortifications excavated in 2008–2010 show evidence of a similar sequence of events, however, with less evidence of burning. The structures were demolished with a large number of *in situ* pottery still in the rooms. These buildings were at least in part reconstructed along the same lines. The High Inner City Temple underwent several phases of rebuilding and modification before its destruction by fire. Three radiocarbon samples, likely the remains of poplar roof beams, were collected from its final floor (Table 1). These provide a date of ca. 2200–2100 BC for the construction of the roof and hence the destruction dates to sometime in the late third millennium BC. There is evidence of building activity and occupation within the temple following the burning, but it does not appear to be more than a squatter occupation—the mudbrick walls were baked by the fire, creating a durable ruin that was slowly buried in the accumulating debris of the early second-millennium settlement. Following Sweyhat Period 4, the

TABLE 1. Radiocarbon Dates from Tell es-Sweyhat and Tell Hajji Ibrahim, Syria

Hajji Ibrahim (HI) Period A1–A2 (EBA-LC Transitional)

AA30469 HI Op. 1 Locus 2 Lot 7 — Charcoal — 4605±55 BP

One Sigma Ranges:	3509–3426 cal BC	Prob. 0.529892
	3382–3335 cal BC	Prob. 0.333049
	3211–3191 cal BC	Prob. 0.078586
	3152–3136 cal BC	Prob. 0.058474
Two Sigma Ranges:	3621–3609 cal BC	Prob. 0.009471
	3521–3309 cal BC	Prob. 0.739361
	3299–3283 cal BC	Prob. 0.013018
	3276–3265 cal BC	Prob. 0.009689
	3240–3104 cal BC	Prob. 0.228461

AA30477 HI Op. 1 Locus 20 Lot 3 — Charcoal — 4435±65 BP

One Sigma Ranges:	3324–3233 cal BC	Prob. 0.33531
	3222–3220 cal BC	Prob. 0.006424
	3172–3161 cal BC	Prob. 0.03716
	3117–3009 cal BC	Prob. 0.464086
	2983–2935 cal BC	Prob. 0.15702
Two Sigma Ranges:	3338–3206 cal BC	Prob. 0.323103
	3195–2919 cal BC	Prob. 0.676897

AA30478 HI Op. 1 Locus 18 Lot 3 — Charcoal — 4425±55 BP

One Sigma Ranges:	3310–3295 cal BC	Prob. 0.051376
	3285–3275 cal BC	Prob. 0.033848
	3265–3239 cal BC	Prob. 0.109523
	3105–2927 cal BC	Prob. 0.805252
Two Sigma Ranges:	3335–3211 cal BC	Prob. 0.279255
	3191–3152 cal BC	Prob. 0.065959
	3137–2915 cal BC	Prob. 0.654786

AA30471 HI Op. 3 Locus 2 Lot 5 — Carbonized Seeds — 4435±55 BP

One Sigma Ranges:	3322–3272 cal BC	Prob. 0.186711
	3269–3235 cal BC	Prob. 0.145425
	3221–3221 cal BC	Prob. 0.003184
	3171–3162 cal BC	Prob. 0.030738
	3116–3010 cal BC	Prob. 0.506509
	2980–2957 cal BC	Prob. 0.080651
	2953–2939 cal BC	Prob. 0.046783
Two Sigma Ranges:	3337–3208 cal BC	Prob. 0.31995
	3193–3149 cal BC	Prob. 0.083084
	3140–2920 cal BC	Prob. 0.596965

continued

TABLE 1. *continued*

AA30476 HI Op. 3 Locus 2 Lot 7		Charcoal	4420±55 BP
One Sigma Ranges:	3308–3301 cal BC	Prob. 0.022932	
	3282–3277 cal BC	Prob. 0.012904	
	3265–3240 cal BC	Prob. 0.09878	
	3104–2924 cal BC	Prob. 0.865385	
Two Sigma Ranges:	3334–3212 cal BC	Prob. 0.25733	
	3190–3153 cal BC	Prob. 0.058149	
	3135–2912 cal BC	Prob. 0.68452	

AA30475 HI Op. 3 Locus 2 Lot 7		Charcoal	4380±65 BP
One Sigma Ranges	3090–3042 cal BC	Prob. 0.242587	
	3039–2912 cal BC	Prob. 0.757413	
Two Sigma Ranges:	3330–3215 cal BC	Prob. 0.153831	
	3185–3156 cal BC	Prob. 0.027744	
	3126–2890 cal BC	Prob. 0.818425	

Sweyhat (SW) Period 1/Hajji Ibrahim Period B (EBI)

AA30472 HI Op. 2 Locus 8 Lot 1		Organic Matter	4405±55 BP
One Sigma Ranges:	3261–3255 cal BC	Prob. 0.027667	
	3097–2921 cal BC	Prob. 0.972333	
Two Sigma Ranges:	3331–3214 cal BC	Prob. 0.196324	
	3186–3156 cal BC	Prob. 0.035549	
	3127–2905 cal BC	Prob. 0.768126	

AA30468 HI Op. 2 Locus 8 Lot 1		Carbonized Seeds	4330±55 BP
One Sigma Ranges:	3012–2947 cal BC	Prob. 0.54272	
	2945–2897 cal BC	Prob. 0.45728	
Two Sigma Ranges:	3263–3247 cal BC	Prob. 0.008362	
	3100–2874 cal BC	Prob. 0.991638	

AA30473 HI Op. 2 Locus 8 Lot 1		Charcoal	4280±55 BP
One Sigma Ranges:	3010–2979 cal BC	Prob. 0.146568	
	2958–2951 cal BC	Prob. 0.026805	
	2941–2871 cal BC	Prob. 0.743103	
	2801–2779 cal BC	Prob. 0.083525	
Two Sigma Ranges:	3085–3063 cal BC	Prob. 0.018132	
	3028–2849 cal BC	Prob. 0.793694	
	2813–2740 cal BC	Prob. 0.142398	
	2729–2693 cal BC	Prob. 0.041948	
	2687–2679 cal BC	Prob. 0.003827	

continued

TABLE 1. *continued*

AA32662 SW Op. 1 Locus 52 Lot 3		Charcoal	4140±65 BP
One Sigma Ranges:	2869–2830 cal BC	Prob. 0.192891	
	2822–2803 cal BC	Prob. 0.089824	
	2778–2628 cal BC	Prob. 0.717285	
Two Sigma Ranges:	2889–2569 cal BC	Prob. 0.983395	
	2515–2501 cal BC	Prob. 0.016605	

Sweyhat Period 2/Hajji Ibrahim Period C (EBII)

AA50639 SW Op. 29 Locus 13 Lot 1		Charcoal	
	4287±47 BP		
One Sigma Ranges:	3008–2985 cal BC	Prob. 0.130815	
	2934–2876 cal BC	Prob. 0.869185	
Two Sigma Ranges:	3079–3070 cal BC	Prob. 0.007615	
	3025–2862 cal BC	Prob. 0.913199	
	2807–2758 cal BC	Prob. 0.070599	
	2718–2707 cal BC	Prob. 0.008587	

AA50641 SW Op. 31 Locus 8 Lot 8		Charcoal	4224±46 BP
One Sigma Ranges:	2901–2860 cal BC	Prob. 0.428764	
	2808–2756 cal BC	Prob. 0.466172	
	2719–2704 cal BC	Prob. 0.105063	
Two Sigma Ranges:	2914–2835 cal BC	Prob. 0.391336	
	2817–2665 cal BC	Prob. 0.601745	
	2645–2639 cal BC	Prob. 0.006919	

AA50640 SW Op. 21 Locus 118 Lot 1		Charcoal	4138±62 BP
One Sigma Ranges:	2868–2829 cal BC	Prob. 0.186865	
	2823–2803 cal BC	Prob. 0.090921	
	2777–2626 cal BC	Prob. 0.722214	
Two Sigma Ranges:	2887–2571 cal BC	Prob. 0.987743	
	2513–2503 cal BC	Prob. 0.012257	

Sweyhat Period 3 (EBIII-IVa)

No radiocarbon dates are currently available

Sweyhat Period 4 (EBIVa-IVb)
Area IV Building:

GrN10350	SW Area IV P1.5	Charcoal	3810±35 BP
One Sigma Ranges:	2297–2198 cal BC	Prob. 0.936545	
	2163–2152 cal BC	Prob. 0.063455	
Two Sigma Ranges:	2453–2446 cal BC	Prob. 0.004264	
	2435–2420 cal BC	Prob. 0.013491	
	2405–2378 cal BC	Prob. 0.032038	
	2349–2138 cal BC	Prob. 0.950207	

continued

TABLE 1. *continued*

P2328	SW Area IV Room 7	Charcoal	3730±70 BP
One Sigma Ranges:	2275–2254 cal BC	Prob. 0.074987	
	2227–2224 cal BC	Prob. 0.009646	
	2209–2028 cal BC	Prob. 0.915367	
Two Sigma Ranges:	2397–2384 cal BC	Prob. 0.006203	
	2346–1931 cal BC	Prob. 0.993797	

GrN10349	SW Area IV F1.15	Charcoal	3675±40 BP
One Sigma Ranges:	2134–2077 cal BC	Prob. 0.506963	
	2073–2070 cal BC	Prob. 0.012695	
	2064–2018 cal BC	Prob. 0.374998	
	1995–1981 cal BC	Prob. 0.105344	
Two Sigma Ranges:	2195–2170 cal BC	Prob. 0.042027	
	2145–1944 cal BC	Prob. 0.957973	

P2324	SW Area IV Building	Carbonized Grain	3640±70 BP
One Sigma Ranges:	2133–2082 cal BC	Prob. 0.242896	
	2059–1919 cal BC	Prob. 0.757104	
Two Sigma Ranges:	2204–1871 cal BC	Prob. 0.958619	
	1845–1812 cal BC	Prob. 0.023791	
	1802–1776 cal BC	Prob. 0.01759	

Temple:

AA50645 SW Op. 34 Locus 101 Lot 10		Charcoal	3797±45 BP
One Sigma Ranges:	2295–2192 cal BC	Prob. 0.772377	
	2179–2142 cal BC	Prob. 0.227623	
Two Sigma Ranges:	2457–2419 cal BC	Prob. 0.032282	
	2407–2376 cal BC	Prob. 0.034478	
	2351–2128 cal BC	Prob. 0.889725	
	2088–2046 cal BC	Prob. 0.043516	

AA50644 SW Op. 34 Locus 102 Lot 4		Charcoal	3745±71 BP
One Sigma Ranges:	2280–2250 cal BC	Prob. 0.121782	
	2230–2219 cal BC	Prob. 0.037869	
	2211–2035 cal BC	Prob. 0.840349	
Two Sigma Ranges:	2451–2447 cal BC	Prob. 0.00259	
	2436–2420 cal BC	Prob. 0.00882	
	2405–2378 cal BC	Prob. 0.017671	
	2349–1947 cal BC	Prob. 0.970919	

AA50643 SW Op. 34 Locus 102 Lot 3		Charcoal	3740±45 BP
One Sigma Ranges:	2204–2122 cal BC	Prob. 0.645756	
	2094–2041 cal BC	Prob. 0.354244	
Two Sigma Ranges:	2289–2025 cal BC	Prob. 1.00	

continued

TABLE 1. *continued*

Other:			
GrN10348	SW Area III B2.6	Charcoal	3675±80 BP
Sweyhat 5–6 (EBA-MBA Transitional)			
	No radiocarbon dates are currently available		
Sweyhat 7 (Seleucid)			
GrN9203 SW Area II 7.2		Charcoal	2165±35 BP
One Sigma Ranges:	353–293 cal BC	Prob. 0.529813	
	230–218 cal BC	Prob. 0.070468	
	213–168 cal BC	Prob. 0.399719	
Two Sigma Ranges:	362–269 cal BC	Prob. 0.446516	
	265–108 cal BC	Prob. 0.553484	

settlement began to shrink until only the area of the High Mound was occupied in late Periods 5 and 6, although this was a gradual process that lasted well into the early second millennium.

What makes the Sweyhat Period 4 urbanization so compelling is that it occurred at the height of the 4.2 ka climatic event that brought centuries of aridification and abandonment to the Khabur region. Tell es-Sweyhat lasted at least another 150–200 years as an urban center and then another 100–200 years as a declining town. Weiss and his colleagues, in their reconstructions of the cultural response to the 4.2 ka event in western Syria, cite the widespread abandonment and decline of a number of sites:

> Similar city and town abandonments [to the Khabur Plains] occurred across western Syria, for instance, Ebla . . . and Umm al-Marra . . . and even along the middle Euphrates at Jerablus Tahtani . . . Selenkahiyeh . . . Halawa . . . and Sweyhat . . . where reduced settlements, buffered by river access, survived (Staubwasser and Weiss 2006, 10–11).
>
> Along the middle Euphrates, the settlement systems dominated by the Tell Banat complex . . . in the Early Dynastic III period quickly collapsed ca. 2200 BC. Massive Euphrates flooding, a product perhaps of depleted vegetation cover and reduced soil infiltration capacity, has been identified for this same period by Tipping and Peltenburg at Jerablus Tahtani . . . During the aridification period, reduced settlement, reduced aggregate population, and reduced individual site sizes were characteristic of this region . . . Further downstream, Tell Sweyhat's changing sizes and patterns of growth appear difficult to define, but the pattern of late third millennium growth and pastoralism . . . coincided with reduction of populations in the dry farming Habur Plains to the north. This conforms to epigraphic and ethnographic data for habitat-tracking within pastoralist movement down the Euphrates during periods of drought . . . (Weiss 2000, 88).

This is an exaggeration. There are no widespread regional abandonments on the order of the Khabur collapse at 4.2 ka in other parts of northern Meso-

potamia, although many settlements in the latter part of the third millennium experienced disruptions followed by rebuilding. The more pronounced break comes in the EBA-MBA transition of approximately 2000–1900 BC, two centuries or more after the onset of the 4.2 ka climate event, and even then there are important continuities between the EBA and the MBA (Cooper 2006, 25–28; Schwartz and Miller 2007, 199). There is no major late third millennium BC break at Ebla, as evidenced by the remains of Period IIB2, until the disruption of around 2000 BC (Dolce 1999, 2002; Pinnock 2009). Umm al-Marra has abundant evidence for occupation during the later third millennium (Schwartz et al. 2000, 450) with a disruption in the EBA-MBA transition, although there is also evidence for continuity at sites in the Jabbul region from EBIVb to the MBA (Schwartz and Miller 2007). Jerablus Tahtani was abandoned around 2200 BC (Peltenburg 1999, 103; Peltenburg et al. 1995, 14–15), but the period from the late third millennium BC to the early second represents a peak period of settlement in the surrounding region (Algaze, Breuninger, and Knutstad 1994, 14–17). At Selenkahiye there is a destruction at the end of EBIVa, followed by rebuilding and another destruction at the end of EBIVb (Van Loon 2001). Many sites in the Middle Euphrates region decline in the late third millennium, including Halawa (Orthmann 1989), Tell Kebir (Porter 1995), and Hadidi (Dornemann 1985), and we see a similar pattern in the Sweyhat region around 2000–1900 BC. Nonetheless, Weiss and his colleagues have conflated the disruptions outside the Khabur and overstated their severity. In fact, the onset of the 4.2 ka event coincides with a peak period of settlement in the Balikh-Euphrates upland steppe (Danti 2000). Similarly, there is no evidence at Tell es-Sweyhat for a decline until a *shift* in settlement patterns begins in the EBA-MBA transitional period, followed by the site's abandonment in the early second millennium BC in favor of a settlement pattern focused on or near the floodplain (Wilkinson 2004, 143). This settlement pattern is, in effect, a return to the "norm" seen in previous periods and is, in general, the typical pattern of settlement until the second period of expansion onto the main terrace and into the high plateau in the Seleucid and Late Roman periods (Wilkinson 2004, 145–47). The question is not whether there was a collapse in western Syria in the late third millennium concurrent with the 4.2 ka event, but rather why the Khabur region specifically experienced widespread urban devolution while other regions prospered until the EBA-MBA transition? Put another way, what factors made the Khabur region more susceptible to the 4.2 ka event and/or made the Middle Euphrates more resilient? Any investigation of the Khabur region must take into account its ties with the Akkadian Empire at the time of the urban collapse.

THE COLLAPSE OF THE AKKADIAN EMPIRE

The link between the collapse of the Akkadian Empire and the 4.2 ka event is primarily based on archaeological evidence from the Khabur Plains and three assumptions: (1) that the 4.2 ka event adversely affected southern Mesopotamia despite the buffering effect of canal irrigation; (2) that the Khabur

region was economically vital to the Akkadian state; and (3) that "collapse" in the Khabur put southern Mesopotamia at risk from incursions of northern refugees.

Southern Mesopotamia and the 4.2 ka Event

Two lines of evidence have been cited for the potential impact of deteriorating climate on southern Mesopotamia. Climate proxy data indicate a period of reduced precipitation in Anatolia, the primary catchment zone of the Tigris and Euphrates, which has been linked to the NAO (Cullen and deMenocal 2000). Deep sea coring in the Gulf of Oman has revealed increased airborne dust, indicative of aridification in southern Mesopotamia and closely linked in time to the Khabur collapse through geochemical analysis of volcanic ash sherds found in archaeological deposits in the Khabur and the sea cores (Cullen et al. 2000). Whether we see an immediate impact in southern Mesopotamia similar to that in the Khabur region is open to interpretation and hinges on our chronology of the late third millennium BC. In general, there are two schools of thought on the aftermath of the Akkadian collapse. One view, placing weight on the historical record, argues for a period of disruption and political destabilization culminating in the Gutian invasion from the east recorded in the Sumerian King List; however, there is no firm archaeological evidence for a "Gutian Interregnum." The Akkadian Empire did not necessarily end precipitously, but rather seems to have gradually shrunk back to the area of its capital, Agade. There may be overlap, or only a brief hiatus, between the last Akkadian kings and the rise of moderately powerful rulers in southern Mesopotamia that set the stage for the Ur III kingdom. In short, while there is political fragmentation, its extent and duration are far from clear. Overall, it appears that there was a fallback to city-states, the normal political pattern in Mesopotamia, and that there was no catastrophic disruption. The rise of the Ur III state contradicts the notion that the 4.2 ka event had a profound and immediate impact on the south. In many ways, we see a similar pattern between northern and southern Mesopotamia: a period of disruption at around 2200 BC followed by fairly rapid recovery and then a major disruption at the end of the third millennium BC.

Economic Relations between the Khabur and the Akkadian Empire

Reconstructing economic interactions between the Akkadian state and its provinces such as the Khabur is hindered by an almost complete lack of evidence from the Akkadian core area. The subsequent period of political regeneration, the Ur III period, which has a surfeit of textual data on economics, may provide some insights. The question is whether the Ur III state provides an ethnohistorical analogy for reconstructing: (1) Akkadian imperialism, or (2) a cultural response to Middle Holocene climatic deterioration, and/or (3) lessons learned from the failings of the Akkadian state.

UR III *Pastoral Finance and Akkadian Imperialism*

The economic relationships between the core area of the Ur III Empire (2112–2004 BC) and its provinces are well documented in the records of the gún ma-da and mas(h)-da-ri-a annual taxes (Steinkeller 1991). These records show that the Ur III kings dominated the piedmont of the Zagros Mountains, an area in which, much like northern Mesopotamia, marginal zone rainfall agriculture and pastoral production formed the basis of the economy (Steinkeller 1991, 31, fig. 6). Annual taxes in the Ur III provinces were largely paid by military personnel to the central bureaucracy in sheep and, to a lesser extent, oxen. Unlike the Middle Euphrates region, this is a zone of vertical transhumance: flocks graze the cooler, wetter uplands in summer to exploit seasonal pastures and in winter are brought to the warmer, lower piedmont and floodplain where conserved fodder (barley straw and grain) is available. In contrast to the Akkadian state, the Ur III kings were less interested in, or were incapable of, controlling the area of "horizontal" transhumance dominated by the independent city-states of the Middle Euphrates and western Syria. This area posed a threat in the form of the so-called Amorites, as presumably did much of the benighted Khabur with the exception of centers such as Tell Brak (see below).

The advantages of the Ur III system of pastoral finance, as opposed to simple staple finance (grain), are readily apparent: pastoral production could be easily centralized since transporting live animals was far easier than moving bulk shipments of grain. The low annual rainfall of the western piedmont of the Zagros (200–400 mm per annum), similar to conditions in much of northern Mesopotamia, selects for the production of barley. Since barley was primarily fodder, it was not likely shipped at exorbitant expense to centralized, state-controlled storage facilities in southern Mesopotamia, where irrigation agriculture provided bountiful harvests of wheat for human consumption. Instead, fodder production was likely managed by local authorities for the support of independent pastoralists and state herds. The state management of conserved fodder provided some control over seasonally transhumant pastoralists, who presented a potential threat to the state's stability. Here we touch on one of the primary flaws of some models of EBA Mesopotamian state economies (and their collapses): the assumption that grain produced in the periphery constituted a foodstuff that was imported to the core. As has been pointed out, southern Mesopotamia's periphery was largely an agriculturally marginal zone characterized by agropastoralism (the rainfall farming of barley and herding of small ruminants). The geographically regular distribution of stored staples, that is, conserved fodder, is a prerequisite of production. Hegemonic states need only to control fodder sources to tap pastoral production—the surplus to be extracted are the live animals and their byproducts (especially wool). State-controlled storage facilities in the periphery would have served this end, not the bulk transport of grain to southern Mesopotamia. A key point is the cost (measured in grain) of transporting bulk commodities. Riverine transshipment of grain from the periphery, while possible, is not widely attested, even in well-documented periods. Long-distance

overland transshipment is scarcely documented at best, and draft animals require a high-energy and protein-rich source of fodder (grain) for such strenuous work. However, these limitations do not preclude a staple-finance model. We need to avoid the assumption that there was a highly centralized system of grain redistribution. Hegemonic states in Mesopotamia were fueled by the control and strategic location of staples. The profits to be collected from the marginal zone were the livestock supported by rainfall agriculture and seasonal pastures. Campaigning armies and defending troops largely resided and maneuvered at and near the periphery, providing another reason for the regularized distribution of staples. Finally, the Ur III state's meticulous concern with incorporating pastoral production into its economy at the supraregional level may also be indicative of a strategy for coping with the ongoing climatic deterioration of the late third millennium BC.

The economic policies of the Ur III state demonstrate the importance of tapping agropastoral production in the marginal agricultural zones of Mesopotamia. Those zones provided a source of readily transportable wealth in the form of herds, perhaps also revealing a cultural response to changing climatic conditions. The main problem is in determining whether a similar economic pattern was typical of the earlier Akkadian Empire. As previously mentioned, the Akkadian textual sources are mute on this point and our archaeological view of the empire is from the outside looking inward. An Akkadian political and economic presence in the Khabur is archaeologically documented, and there is little need to repeat this oft-cited evidence (Weiss et al. 1993; Michalowski 1993). In western Syria, cities the Akkadian rulers claim to have defeated and that provide corroborating evidence of destruction, such as Ebla, lack an Akkadian administrative presence in the EBIVa. Ebla shows evidence of rebuilding and prosperity following the EBIVa destruction of some of its monumental buildings, a disruption often linked to Sargon or Naram-Sin, but now thought to predate the Akkadian dynasty (Archi and Biga 2003, 29–35). Tell es-Sweyhat, like many other sites, experienced destruction of some of its monumental urban core in the later third millennium, but the radiocarbon dates currently available (Table 1) place this event(s) later than the political and military ascendancy of Akkad, although admittedly more absolute dates are needed from Sweyhat Periods 3–4. Other sites in the Middle Euphrates region also show signs of disruption and a few abandonments. If any relation existed between the Middle Euphrates and Akkad, the dataset is more in keeping with predatory raiding and vassal and tributary states than with direct hegemonic rule, although it is still difficult to connect the major EBA sites of the Middle Euphrates to contemporary toponyms (Otto 2006). While the Akkadian presence in the Khabur is indisputable, less clear is the economic link between the two areas. Weiss and his colleagues have argued for a staple-finance model in which the Khabur provided critical grain exports to the Akkadian state (Weiss et al. 1993). While such a relationship is plausible, and the Khabur region's capacity for grain production is widely recognized, Weiss's argument for grain traveling southward to support the Akkadian state is highly unlikely and virtually undocumented (Zettler 2003, 20–21 n. 8). Still, the question remains: if agro-

pastoralism emphasizing surplus pastoral production typified the Khabur, why was it so prone to collapse?

As one moves away from the Khabur, our archaeological dataset from the late third millennium BC reveals a continuation of settled life with some evidence for disruptions until the EBA-MBA transition. Balanced agropastoral production surely provided an important means for withstanding aridification. Again, why might the Khabur region have been so susceptible to climate-driven disruption? To answer this question, we can begin by looking at the effects of three interrelated variables: the effects of Akkadian imperialism, settlement patterns, and agricultural intensification. If the Akkadians were drawing off surplus production from the Khabur, whether pastoral or agricultural, this would have lowered the resilience of the subsistence economy to meet local needs, contributing to instability. Elite mismanagement could have exacerbated such a situation. Another clue to the Khabur's susceptibility and the resilience of other regions might stem from settlement patterns. The nature of the urban landscape of the Middle Euphrates promoted a more balanced mode of agropastoral production, while in the Khabur region the growth of agricultural territories and population likely selected for a higher degree of separation between agriculture and pastoralism. There is a marked difference in the scale of urban settlement between the Khabur and the Middle Euphrates (cf. Akkermans and Schwartz 2003, 246; Bunnens 2007). Unlike the eastern Khabur Plain in the Leilan IIB period, which was characterized by a tightly packed settlement pattern with abutting agricultural territories, the small valleys adjacent to the Euphrates provided isolated tracts of arable land that were hemmed in by adjacent steppe. This hinterland, while providing small pockets of widely dispersed farm ground, was more valuable for its pasture and wild game. During episodes of marked urban expansion, there was always pasture available in the Balikh-Euphrates uplands. In the Khabur and other areas, as sites increased in number and size during periods of urbanization, agricultural territories immediately abutted each other (Wilkinson 1994, 492–93), seasonally pushing pastoralists off the higher rainfall portions of the plain and moving pastoral production toward a specialist activity outside the sociopolitical sphere of cities. In the Khabur, spring pastures were further removed from sources of conserved winter fodder. An Akkadian imperial presence in the Khabur would also likely have motivated herders to alter their seasonal transhumance if politically and economically expedient. Again, elite mismanagement could easily worsen such a precarious situation. In summary, socioeconomic divergence was a key obstacle faced by expanding states shifting from smaller-scale regional economies with highly integrated, regionally based agricultural and pastoral production to imperial economies characterized by a higher degree of regional specialization and intensification in agropastoralism. Specialization, intensification, and bureaucratic (mis)management promoted the separation of pastoral and agricultural production, thereby facilitating the production of surpluses at the expense of long-term resilience and stability. The weak link in this system was pastoral production: peasant farmers have far fewer options when faced with instabilities, but pastoralists adapt through transhumance and thus vital

economic relations are severed. The rift deepens as pastoralists become socio-politically independent.

Invaders from the North

Scholars have linked the 4.2 ka event to widespread population displacements in northern Mesopotamia, key among them the Amorites who contributed to the collapse of late third millennium empires in southern Mesopotamia (Weiss et al. 1993). In earlier analyses, the Amorites were seen as an important force in disrupting southern Mesopotamian civilization (Weiss et al. 1993). This theory required the expedient transformation of Subartians into Amorites (Weiss et al. 1993, fig. 1) and the fortuitous localization of the ambiguous Guti in southeastern Anatolia. More recent reworkings of the Catastrophic Model use the ill-defined concept of tribal pastoralist "habit tracking" (Weiss and Bradley 2001, 61) to explicate both long-distance population displacements and MBA settlement patterns in the Khabur (Ristvet and Weiss 2005, 11). In another usage, "habit-tracking" seems to serve as a direct substitute for the term "transhumance" (Ristvet and Weiss 2000, 94) while in others it is more akin to migration due to drought (Weiss 2000, 88). In terms of scale, "habit-tracking" as applied in the Catastrophic Model seemingly covers the seasonal movement of transhumant pastoralists and, although unattested in the EBA-MBA of northern Mesopotamia, nomads, not to mention the movements of displaced populations. In general terms, the notion of habit-tracking seems to serve as a means for causally linking the collapse of urban settlement in the Khabur at the onset of the 4.2 ka event to larger disruptions of the EBA-MBA transitional period by "Amoritizing" putative displaced Subartians, who then later resettle the Khabur in Leilan Period I. It is important to note here that the continuities between the EBA and MBA are most readily apparent in the area of the Middle Euphrates and western Syria, not the Khabur.

In the aftermath of the Khabur collapse, the local population would likely have reorganized local production to emphasize resilience and stability; that is, smaller scale settlement and heightened emphasis on pastoral production with increased transhumance, a marked separation between agricultural and pastoral producers, and a reversion to lower-order sociopolitical units. There is no evidence of mass migrations from the Khabur—if the attestations of ambiguous exonyms in southern Mesopotamian texts tell us anything, we read of increasing problems with Amorites in the south, not Subartians.[13] The decline of an urban landscape, as documented by archaeological survey, does not necessarily entail a decline in cultivation. Too often collapse is equated with increases in the nonsedentary pastoralist population—in the case of the 4.2 ka climate event, wandering "Amorites." I would counter that sheep-goat pastoralism in all its forms practiced in northern Mesopotamia required a ready supply of cold-season conserved fodder provided by the dry farming of barley. In the Khabur, the decline in urbanism and the removal of Akkadian rule likely *facilitated* such a production strategy following an initial disruption.

The Late Holocene: Increasing Aridity, Transhumant Pastoralism, and Nomadism

Following the 4.2 ka event, climate improved, but overall a pattern of increasing aridity prevailed in the Late Holocene (Bar-Matthews, Ayalon, and Kaufman 1998; Lemcke and Sturm 1997, fig. 5). In the early second millennium BC, Tell es-Sweyhat slowly declined. The settlement was confined to the High Mound, and the city's inner fortifications had fallen out of use with houses built atop the old foundations. During the EBA-MBA transition, settlement in the region was confined to the area of the floodplain and the main terrace, with the adjacent uplands apparently devoid of settlement. This does not represent a collapse; rather, there was a shift to the vicinity of the floodplain with new and smaller walled towns attested in the MBII period. This reflects the end product of a gradual process of reorganization started in the late third millennium. With increased aridity in the late third and early second millennia, there was a gradual shift from intensive, urban-based agropastoralism to a more resilient strategy based on a higher degree of pastoral transhumance characterized by increased socioeconomic separation between farmers and herders. This is supported by historical documentation from the MBA for the Middle Euphrates region, especially the Mari Letters, which indicate that transhumant pastoralists comprised a significant component of the population. Thus we see the development of a dimorphic form of agropastoralism characterized by greater socioeconomic separation between pastoral specialists and agriculturalists.

Unlike the Khabur, there is no MBII urban recovery in the Tell es-Sweyhat region, or, relative to the EBA, for the southern Middle Euphrates region as a whole. Dense, highly nucleated and hierarchically organized settlement systems did not return to the areas away from the Euphrates. Settlement shrank to the area of the floodplain in the MBA, and the main terrace and uplands were not extensively resettled until the Seleucid and Late Roman periods. In this regard, the EBA pattern of main terrace settlement stands out as an anomaly.

Why did dense settlement not return to some areas as the aridification event abated? Climate-driven catastrophes alone do not account for such patterns. Instead, we should look at the effects of longer-term climate change in the Middle and Late Holocene, environmental degradation, and the interplay between sociopolitical organization and subsistence economies. The peaks in settlement of the EBA and later Seleucid–Late Roman periods occurred during episodes with strong polities promoting agropastoral production. Periods with more predatory political overlords see the destabilization of sedentary settlement and poor relations with transhumant and nomadic pastoralists.

In the Late Bronze Age (LBA), settlement patterns were quite similar to those in the MBA. In the sphere of agropastoral production, we see a "nomadic revolution" in northern Mesopotamia and western Syria. The period witnessed the rise of Aramaean states in the region and the emergence of true

nomadism with the introduction of the domesticated camel and the realization of the capabilities of the horse. The camel made long-range seasonal migrations into the desert steppe possible, greatly altering patterns of seasonal transhumance across the Near East, trade routes, regional politics, and settlement patterns. The settled populace was now confronted by a highly mobile, and potentially hostile, military power. The seasonal movements of transhumant "sheep tribes" were no longer dictated by the vagaries of rainfall and shifting tribal territories and alliances. Herders now had to contend with the seasonal arrival of waves of nomads.

Climate seems to have trended toward increasing aridity in the Iron Age. The Sweyhat region was almost exclusively used for nomadic and transhumant pastoralism with little to no settlement away from the river in the early Iron Age. The notable exception is the Seleucid to the Abassid periods. Although we are in the process of refining our archaeological chronology of these periods in the Middle Euphrates, overall we see a settlement pattern similar to that of the Early Bronze Age. Once again sedentary occupation extended onto the main terrace and into the surrounding uplands (Danti 2000; Wilkinson 2004), although the sites are far more modest in comparison to those in the late third millennium BC. Interestingly, this spike in the number of sites and total occupied area was coterminous with a marked phase of aridity, evincing a trend seen previously with the 5.2 ka and 4.2 ka climate events. While developments in the Late Holocene, particularly the rise of nomadism and major shifts in settlement patterns, are especially interesting, much more archaeological research on this period is needed in the Middle Euphrates region. In particular, we must refine our archaeological chronology to match the resolution of paleoclimatology through the excavation of Iron Age sites, especially smaller sites in rural hinterlands.

CONCLUSIONS

In northern Mesopotamia, Middle and Late Holocene climate events do not appear to have caused punctuated, monolithic responses because northern Mesopotamia was not a homogeneous cultural or environmental region. That said, in the mid- to late Holocene shorter-term periods of pronounced aridity and a long-term trend toward increasing aridity profoundly influenced subsistence economies in the region, providing harsh tests of the resilience and stability of agropastoral systems. Overall, until the modern era, the trend was to emphasize the pastoral component of agropastoral economies by increasing mobility as environmental conditions deteriorated due to anthropogenic environmental degradation, climatic fluctuations, and climate change. The major stages in this process were, first, intensive agropastoralism, then dimorphic agropastoralism, and finally nomadism. This developmental trajectory was most apparent in the marginal zone of northern Mesopotamia, where climate was always a major challenge and the risks inherent in dry farming were offset by pastoral production.

As research has continued on the cultural responses to climate events in northern Mesopotamia, building on the influential work of researchers such as Harvey Weiss, Tony Wilkinson, and Frank Hole, among others, one of the more important developments has been a change in our understanding of the timing and uneven impact of these events, which do not necessarily correspond with widespread disruptions in settlement. If anything, there seems to have been a lag in cultural response. In the case of the 4.2 ka event, certainly the most widely studied climate event after the Younger Dryas, there seems to have been no sudden impact in northern Mesopotamia outside of the collapse documented in the Khabur Plain. This is not the same as saying there was no cultural response, as we see prosperity in areas with smaller scale states and balanced agropastoral economies. Increasingly, the archaeologist's attention has been shifting from the period of the Akkadian Empire to the EBA-MBA transition, which seems to correspond with the final part of the climate event. The EBA-MBA transition, as originally defined, marks a major cultural disruption; change is manifest in the archaeological record, but there is also cultural continuity from the perspective of the Middle Euphrates and western Syria.

The "Sweyhat paradigm," the extensive EBA settlement pattern in the marginal zone of the Middle Euphrates, would not be replicated until the modern era of mechanized farming, pump irrigation, and forced sedentarization of transhumant pastoralists and nomads. Similar phenomena have been documented elsewhere, such as around Tell al-Hawa. Wilkinson attributes this to anthropogenic environmental degradation:

> There followed [the MBA] a steady decline in settlement until the plain was virtually deserted by the 18th/19th century AD . . . Such a long-term decline suggests that after the 2nd Millennium BC settlement, although viable, could not occur on the scale that existed during the 4th, 3rd and early 2nd Millennium . . . If intensive cultivation was continued for sustained periods over such an area it is likely that soil deterioration would also have resulted. Periodic droughts or longer spells may have contributed a significant amount of dust to the atmosphere from the rain-fed zone, thereby exacerbating soil degradation and potentially influencing local climate (1997, 98–99).

In addition to human impact on the environment, long-term climate change has been implicated (Courty 1994; Frumkin et al. 1994). While researchers have targeted intensive agricultural production as a major source of anthropogenic environmental degradation, more attention should be given to the impact of pastoral production. Certainly, the overgrazing of steppic pastures, particularly in regions prone to erosion such as the Balikh-Euphrates uplands, is linked to the decline of EBA intensive agropastoral production and the related settlement patterns. These marginal and delicate environmental zones were far more vulnerable, especially during punctuated climate events. In the case of the Middle Euphrates, during the mid-third millennium

BC, a period of relatively moist climatic conditions, intensive agropastoral production flourished. This subsistence strategy provided a means for maximizing production by linking the pockets of arable land along the Euphrates to the adjacent upland steppic pastures. It required minimal seasonal transhumance and thus could be fully integrated within individual sociopolitical and economic units. This strategy emphasized resilience, providing a means for coping with pronounced interannual variation in rainfall. In the late third millennium, we see the maximization of this system during a period of heightened aridity. The weak link in such a system would have been steppic pastures and the destructive force of overgrazing. Declines in the quality of rangeland and the increasing aridity of the Late Holocene in the Middle Euphrates region would have resulted in lower stocking rates on steppic pastures and the attendant response would have been increased transhumance and the emergence of dimorphic agropastoral production in the region—the pattern documented in the MBA and later. In the Late Holocene, modest settlements lined the Euphrates floodplain, and the surrounding steppe was used almost exclusively as rangeland by transhumant and nomadic pastoralists. Further research is needed to test this in the Middle Euphrates region and in other areas, particularly in regions of the transitional zone of Syria that experienced similar long-term patterns. Climate and environmental degradation are prime driving forces operating on a global scale, but in our search for cultural responses we must never forget the "butterfly effect"—small variations in the initial conditions of a dynamical system may produce large variations over the *longue durée*.

NOTES

1. This inherent contradiction has been remarked upon by Brooks (2006, 45): ". . . it should be noted that the climatic changes apparently associated with the emergence of complex societies are qualitatively if not quantitatively similar to those blamed for societal collapse. . . . In Mesopotamia, aridity appears to be associated with the emergence of the Uruk culture, its subsequent collapse and the transition to the Dynastic period, and also with the later collapse of the Akkadian Empire."

2. One of the main modifications of Wilkinson's model has been his acknowledgement of the importance of pastoral production in the marginal zone of Syria.

3. Recent assessments of the 4.2 ka event are substantially more detailed than earlier portrayals of it (Weiss et al. 1993) as a centuries-long, homogeneous phenomenon. While documented globally, it seems to have only been abrupt in the Near East (Arz et al. 2006). Current reconstructions see its onset at around 2200 BC. In the Middle East, it seems to have been a punctuated, severe drought with a 300-year period of aridity (Cullen et al. 2000), part of a longer-term trend toward increasing aridity (Enzel et al. 2003; Arz, Lamy, and Pätzold 2006).

4. I strongly disagree with Holland's recent re-dating of Sweyhat's urban apogee to the "Akkadian Period," which ignores the published radiocarbon

sequence from the site and more recent findings. This re-dating is part of Holland's argument that some buildings of the Inner City might have been destroyed by Sargon in an attack, although there is no historical documentation to support this.

5. Tell es-Sweyhat was first excavated as part of the Tabqa Dam salvage project in the early 1970s (Freedman 1979; Holland 1976, 1977, 2006). Excavations were resumed in 1989 by the University of Pennsylvania Museum and the Oriental Institute, University of Chicago (Zettler et al. 1997). In 2008, a joint Boston University/University of Pennsylvania Museum project was initiated.

6. Holland (2006, 383) incorrectly identifies Tell Hajji Ibrahim as an EBIII settlement, citing Wilkinson (Wilkinson 2004, 200, Site SS3); however, Wilkinson correctly dates the main occupation of the site to the early EBA. The confusion stems from results published following the initial excavation season at Hajji Ibrahim in which pottery vessels from a grave were attributed to the final phase of occupation at the site (Danti 1996, 1997; Danti and Zettler 1998). In fact, after its abandonment in the early third millennium BC the site was used as a cemetery (Danti 2000, 141–42, figs. 4.13, 5.63, 5.64).

7. Three seasons of excavations were conducted at Tell Hajji Ibrahim under the direction of the author in 1993, 1995, and 1998. Excavations were sponsored by the National Science Foundation, the American Schools of Oriental Research, and P. E. MacAllister.

8. Holland has designated some material from Sweyhat Trench IC as Late Chalcolithic (Holland 2006, 109, 379, fig. 16) based on its stratigraphic position beneath an occupational deposit of EB I/Sweyhat Period 1/Hajji Ibrahim B date (Holland 2006, 380, fig. 17). The material was found in a storage pit cut into sterile soil but cannot be independently dated by style or radiocarbon.

9. Animals can require a 60% increase in digestible nutrients.

10. These dimensions are based on exposures of the structure on the west mound and the south mound. We have yet to reach the levels on which the earliest phase of the building(s) was founded.

11. Tombs have been excavated on the northwestern Low Mound. In 2008 and 2010, irrigation water in the cotton fields located on the Low Mound at Sweyhat opened new tombs in the northern, eastern, and southern Low Mound. Preliminary inspection indicates they are of similar date to those previously excavated.

12. To form the High Inner City, the earlier fortress was completely encased within a new High Terrace, the central focus of the city. The terrace involved the construction of a stone retaining wall, presumably around the area of the fortress, although we need further excavated exposures to be certain it completely encircled it. Structures and earlier, sloping archaeological deposits along the retaining wall's outer periphery were demolished and the debris was used as terrace fill. The Upper Terrace was paved in mudbrick. This effectively created two level surfaces at the center of the settlement: a High Inner City and a Low Inner City.

13. In a seeming attempt to avoid this problem, some scholars posit "habit-tracking" from the Khabur southward down the Euphrates, thus putting displaced Khabur populations in the geographic, and presumably "ethnic,"

position of Amorites (e.g., Weiss 2000, 88; Weiss et al. 1993, fig. 1). These same researchers imply that transhumant pastoralism was not practiced in the third millennium Khabur, thus precluding putative "habits" (Ristvet and Weiss 2005, 11). With regard to Amorite origins, like many researchers I would localize them in the vicinity of the "Big Bend" of the Euphrates, based on continuities between the late EBA and EBA-MBA transition of this region and the later MBA Amorite kingdoms as evinced in ceramics, the prevalence of "family" and communal burials, the emphasis on corporate groups, the strong pastoral component of the economy, and methods of fortification.

REFERENCES

Adams, R. McC. 1978. Strategies of maximization, stability, and resilience in Mesopotamian society, settlement, and agriculture. *Proceedings of the American Philosophical Society* 122: 329–35.

Akkermans, P. M., M. G., and G. M. Schwartz. 2003. *The archaeology of Syria.* Cambridge: Cambridge University Press.

Al-Ashram, M. 1990. Agricultural labor and technological change in the Syrian Arab Republic. In *Labor and rainfed agriculture in West Asia and North Africa,* ed. D. Tully, 163–84. Drodrecht: Kluwer Academic Publishers.

Algaze, G., R. Breuninger, and J. Knudstad. 1994. The Tigris-Euphrates Archaeological Reconnaissance Project: final report of the Birecik and Carchemish Dam Survey Areas. *Anatolica* 20: 1–96.

Archi, A., and M. G. Biga. 2003. A victory over Mari and the Fall of Ebla. *Journal of Cuneiform Studies* 55: 1–44.

Armstrong, J. A., and R. L. Zettler. 1997. Excavations on the high mound (Inner Town). In *Subsistence and settlement in a marginal environment: Tell es-Sweyhat, 1989–1995 preliminary report,* ed. R. L. Zettler et al., 11–34. MASCA Research Papers in Science and Archaeology 14. Philadelphia: Museum Applied Science Center for Archaeology.

Arz, H. W., F. Lamy, and J. Pätzold. 2006. A pronounced dry event recorded around 4.2 kyr in brine sediments from the Northern Red Sea. *Quaternary Research* 66 (3): 432–41.

Bar-Matthews, M., and A. Ayalon. 1997. Late quaternary paleoclimate in the eastern Mediterranean region from stable isotope analysis of speleothems at Soreq Cave, Israel. *Quaternary Research* 47 (2): 155–68.

———. 2004. Speleothems as paleoclimate indicators, a case study from Soreq Cave located in the eastern Mediterranean region, Israel. In *Past climate variability through Europe and Africa,* ed. R. W. Battarbee, F. Gasse, and C. E. Stickley, 363–91. Developments in Paleoenvironmental Research 6. Dordrecht, The Netherlands: Springer.

Bar-Matthews, M., A. Ayalon, and A. Kaufman. 1998. Middle to late Holocene paleoclimate in the eastern Mediterranean region from stable isotopic composition of speleothems from Soreq Cave, Israel. In *Water, environment and society in times of climatic change,* ed. A. S. Issar and N. Brown, 203–14. Amsterdam: Kluwer Academic Press.

Bar-Matthews, M., A. Ayalon, A. Kaufman, and G. Wasserburg. 1999. The eastern Mediterranean paleoclimate as a reflection of regional events: Soreq Cave, Israel. *Earth and Planetary Science Letters* 166: 85–95.

Besançon, J., and P. Sanlaville. 1985. Le milieu géographique. In *Holocene settlement in North Syria*, ed. P. Sanlaville, 7–40. Oxford: BAR.

Bottema, S. 1989. Notes on the prehistoric environment of the Syrian Dezireh. In *To the Euphrates and beyond: Archaeological studies in honour of Maurits N. van Loon*, ed. O. M. C. Haex, H. H. Curvers, and P. M. M. G. Akkermans, 1–16. Rotterdam: A. A. Balkema.

Brooks, N. 2006. Cultural responses to aridity in the middle Holocene and increased social complexity. *Quaternary International* 151: 29–49.

Bunnens, G. 2007. Site hierarchy in the Tishrin Dam and third millennium geopolitics in Northern Syria. In *Euphrates River valley settlement*, ed. E. Peltenburg, 43–54. Levant Supplementary Series 5. Oxford: Oxbow.

Butzer, K. W. 1995. Environmental change in the Near East and human impact on the land. In *Civilizations of the ancient Near East*, vol. 1, ed. J. M. Sasson, 123–51. New York: Scribner.

Capper, B. S., M. Mekni, S. Rihawi, E. F. Thomson, and G. Jenkins. 1985a. Observations on barley straw quality. In *Proceedings of the winter meeting of the British society of animal production*, Scarborough.

Capper, B. S., S. Rihawi, M. Mekni, and E. F. Thomson. 1985b. Factors affecting the nutritive value of barley straw for Awassi sheep. In *Proceedings of the 3rd Animal Science Congress*, Asian-Australian Association of Animal Production Societies. Seoul, Korea.

Charles, H. 1939. *Tribus moutonnières du Moyen Euphrate*. Documents d'Ètudes Orientales VIII. Beirut: Institut Français de Damas.

Cocks, P. S. 1986. Integration of cereal-livestock production in the farming systems of north Syria. In *Potential of forage legumes in farming systems of sub-saharan Africa*, ed. I. Haque, S. Jutzi, and P. J. H. Neate, 186–211. Addis Ababa: International Livestock Centre for Africa.

Cooper, E. 2006. The demise and regeneration of Bronze Age urban centers in the Euphrates Valley of Syria. In *After collapse: The regeneration of complex societies*, ed. G. M. Schwartz and J. J. Nichols, 18–37. Tucson: The University of Arizona Press.

Courty, M.-A. 1994. Le cadre paléogéographique des occupations humaines dans le basin du Haut-Khabur (Syrie du nord-est): Premier resultants. *Paléorient* 20 (1): 21–59.

Cullen, H. M., and P. B. deMenocal. 2000. North Atlantic influence on Tigris-Euphrates streamflow. *International Journal of Climatology* 20: 853–63.

Cullen, H. M., P. B. deMenocal, S. Hemming, G. Hemming, F. H. Brown, T. Guilderson, and F. Sirocko. 2000. Climate change and the collapse of the Akkadian empire: Evidence from the deep sea. *Geology* 28: 379–82.

D'Altroy, T., and T. K. Earle 1985. Staple finance, wealth finance, and storage in the Inca political economy. *Current Anthropology* 26: 187–206.

Danti, M. D. 1996. The Tell es-Sweyhat Regional Archaeological Project. In *Tell es-Sweyhat, 1989–1995: A city in northern Mesopotamia in the third millennium B.C.*, ed. R. L. Zettler, 14–36. *Expedition* 38 (1): 14–36.

————. 1997. Regional surveys and excavations. In *Subsistence and settlement in a marginal environment: Tell es-Sweyhat, 1989–1995 preliminary report,* ed. R. L. Zettler et al., 85–94. MASCA Research Papers in Science and Archaeology 14. Philadelphia: Museum Applied Science Center for Archaeology.

————. 2000. Early Bronze Age settlement and land use in the Tell es-Sweyhat region, Syria. Ph.D. diss., University of Pennsylvania.

Danti, M. D., and R. L. Zettler. 1998. The evolution of the Tell es-Sweyhat (Syria) settlement system in the 3rd millennium B.C. In *Espace naturel, espace habité in Syrie du Nord (10e–2e millenaires av. J.-C.),* ed. M. Fortin and O. Aurenche, 209–28. Bulletin of the Canadian Society for Mesopotamian Studies 33. Lyons: Maison de'l Orient.

deHeinzelin, J. 1967. Investigations on the terraces of the Middle Euphrates. In *The Tabqa Dam reservoir survey,* ed. M. van Loon, 22–27. Damascus: Direction Générale des Antiquités et des Musées.

d'Hont, O. 1994. *Vie Quotidienne des 'Agedat.* Damascus: L'Institut Français D'Études Arabes de Damas.

deMenocal, P. B. 2001. Cultural responses to climate change during the Late Holocene. *Science* 292: 667–73.

Dolce, R. 1999. The 'second Ebla': A view on the EB IVB city. *ISIMU* 2: 293–304.

————. 2002. Ebla after the 'fall'— some preliminary considerations on the EB IVB city. *Damaszner Mitteilungen* 13: 11–28.

Dornemann, R. 1985. Salvage excavations at Tell Hadidi in the Euphrates river valley. *Biblical Archaeologist* 48: 49–59.

Drysdale, R., G. Zanchetta, J. Hellstrom, R. Maas, A. Fallick, M. Pickett, I. Cartwright, and L. Piccini. 2006. Late Holocene drought responsible for the collapse of Old World civilizations is recorded in an Italian cave flowstone. *Geology* 34: 101–4.

Enzel, Y., R. Bookman, D. Sharon, H. Gvirtzman, U. Dayan, B. Ziv, and M. Stein. 2003. Late Holocene climates of the Near East deduced from Dead Sea level variations and modern regional winter rainfall. *Quaternary Research* 60: 263–73.

Epstein, H. 1985. *The Awassi sheep with special reference to the improved dairy type.* FAO Animal Production and Health Paper 57. Rome: Food and Agriculture Organization of the United Nations.

Freedman, D. N. 1979. *Archaeological reports from the Tabqa Dam Project-Euphrates valley, Syria.* Annual of the American Schools of Oriental Research 44. Cambridge, Mass.: American Schools of Oriental Research.

Frumkin, A., I. Carmi, I. Zak, and M. Margaritz. 1994. Middle Holocene environmental change determined from Salt Caves of Mount Sedom, Israel. In *Late Quaternary chronology and paleoclimates of the Eastern Mediterranean,* ed. R. Kra, 315–32. Tucson: University of Arizona.

Ghosh, P. K., and R. K. Abichandani. 1981. *Water and the eco-physiology of desert sheep.* Jodhpur: Central Arid Zone Research Institute.

Goodchild, A. V., A. I. El-Awad, and O. Gürsoy. 1999. Effect of feeding level in late pregnancy and early lactation and fibre level in mid lactation on body mass, milk production and quality in Awassi ewes. *Animal Science* 68: 231–41.

Hole, F. 1991. Middle Khabur settlement and agriculture in the Ninevite V period. *The Canadian Society for Mesopotamian Studies Bulletin* 21: 17–30.

——. 1994. Environmental instabilities and urban origins. In *Chiefdoms and early states in the Near East: The organizational dynamics of complexity,* ed. Gil J. Stein and Mitchell Rothman, 121–51. Monographs in World Archaeology 18. Madison: Prehistory Press.

——. 1999. Storage structures at Tell Ziyadeh, Syria. *Journal of Field Archaeology* 26: 269–83.

Holland, T. A. 1976. Preliminary report on excavations at Tell es-Sweyhat, Syria, 1973–74. *Levant* 8: 36–70.

——. 1977. Preliminary report on excavations at Tell es-Sweyhat, Syria, 1975. *Levant* 9: 36–65.

——. 2006. *Excavations at Tell es-Sweyhat, Syria 2. Archaeology of the Bronze Age, Hellenistic, and Roman remains at an ancient town on the Euphrates River.* 2 vols. Oriental Institute Publication 125. Chicago: The Oriental Institute of the University of Chicago.

Issar, A. S., and M. Zohar. 2004. *Climate change: Environment and civilization in the Middle East.* New York: Springer.

Kaufman, A., M. Bar-Matthews, A. Ayalon, and I. Carmi. 2003. The vadose flow above Soreq Cave, Israel: A tritium study of the cave waters. *Journal of Hydrology* 273: 155–63.

Kuzucuğlu, C., and C. Marro. 2007. Northern Syria and Upper Mesopotamia at the end of the third millennium B.C.: Did a crisis take place? In *Sociétés humaines et changement climatique à la fin du troisième millénaire: une crise a-t-elle eu lieu en Haute Mésopotamie? Actes du Colloque de Lyon, 5–8 décembre 2005.* ed. C. Kuzucuğlu and C. Marro, 583–90. Varia Anatolica XIX. Paris: De Boccard Édition-Diffusion.

Lemcke, G., and M. Sturm. 1997. $d^{18}o$ and trace element measurements as proxy for the reconstruction of climate changes at Lake Van (Turkey): Preliminary results. In *Third millennium BC climate change and Old World collapse,* ed. H. Nüzhet Dalfes, G. Kukla, and H. Weiss, 653–78. NATO ASI Series I, no. 49. New York: Springer.

McCorriston, J. 1992. The Halaf environment and human activities in the Khabur drainage, Syria. *Journal of Field Archaeology* 19: 315–33.

——. 1995. Preliminary archaeobotanical analysis in the Middle Khabur Valley, Syria and studies of socioeconomic change in the early third millennium BC. *Canadian Society for Mesopotamian Studies Bulletin* 29: 33–46.

Michalowski, P. 1993. Memory and deed: the historiography of the political expansion of the Akkad state. In *Akkad: the first world empire: structure, ideology, traditions,* ed. M. Liverani, 69–90. Padova: Sargon.

Miller, N. F. 1990. Clearing land for Farmland and fuel in the ancient Near East. *MASCA Research Papers in Science and Archaeology,* Supplement to Volume 7: 70–78.

——. 1997a. Farming and herding along the Euphrates: environmental constraint and cultural choice (fourth to second millennia B.C.). In *Subsistence and settlement in a marginal environment: Tell es-Sweyhat, 1989–1995 preliminary report,* ed. R. L. Zettler et al., 123–32. MASCA Research Papers

in Science and Archaeology 14. Philadelphia: Museum Applied Science Center for Archaeology.

————. 1997b. Sweyhat and Hajji Ibrahim: some archaeobotanical samples from the 1991 and 1993 seasons. In *Subsistence and settlement in a marginal environment: Tell es-Sweyhat, 1989–1995 preliminary report*, ed. R. L. Zettler et al., 95–122. MASCA Research Papers in Science and Archaeology 14. Philadelphia: Museum Applied Science Center for Archaeology.

Moore, A., and G. Hillman. 1992. The Pleistocene to Holocene transition and human economy in Southwest Asia: The impact of the Younger Dryas. *American Antiquity* 57: 482–94.

Nygaard, D. 1983. Tests on farmers' fields: the ICARDA experience. *Proceedings of the 1st Farming Systems Research Symposium*, Kansas State University, Manhattan, Kansas, 76–98.

Oppenheim, M. F. von. 1939. *Die Beduinen*. Volume 1. Leipzig: Otto Harrassowitz.

Orthmann, W. 1989. *Halawa 1980–1986*. Bonn: Dr. Rudolf Habelt Verlag.

Otto, A. 2006. Archaeological perspectives on the localization of Naram-Sin's Armanum. *Journal of Cuneiform Studies* 58: 1–26.

Peltenburg, E. J. 1999. Jerablus-Tahtani 1992–6: a summary. In *Archaeology of the upper Syrian Euphrates, the Tishrin Dam area*, ed. G. del Olmo Lete and J.-L. Montero Fenollós, 97–105. Aula Orientalis-Supplementa 15. Barcelona: Editorial AUSA.

Peltenburg, E. J., S. Campbell, P. Croft, D. Lunt, M. Murray, and M. Watt. 1995. Jerablus-Tahtani, Syria, 1992–4: Preliminary Report. *Levant* 27: 1–28.

Perrin de Brichambaut, G., and C. C. Wallén. 1963. A study of agroclimatology in the Near East. World Meteorological Organization Technical Note 56. Geneva: World Meteorological Organization.

Pinnock, F. 2009. EB IVB—MBI in northern Syria: Crisis and change of a mature urban civilisation. In *The Levant in transition*, ed. P. J. Parr. Palestine Exploration Fund Annual 9. Leeds, U.K.: Maney.

Porter, A. 1995. The third millennium settlement complex at Tell Banat: Tell Kebir. *Damaszener Mitteilungen* 8: 1–50.

Rihawi, S., B. S. Capper, A. E. Osman, and E. F. Thomson. 1987. Effects of crop maturity, weather conditions and cutting height on yield, harvesting losses and nutritive value of cereal-legume mixtures grown for hay production. *Experimental Agriculture* 23: 451–59.

Ristvet, L., and H. Weiss. 2000. Imperial responses to environmental dynamics at late third millennium Tell Leilan. *Orient Express* (Paris) 2000: 94–99.

————. 2005. The Habur region in the late third and early second millennium B.C. In *The history and archaeology of Syria*, ed. W. Orthmann, 1–26. Vol. 1. Saabrucken: Saarbrucken Verlag.

Roberts, N., and H. E. Wright, Jr. 1993. Vegetational, lake-level, and climatic history of the Near East and South-west Asia. In *Global climates since the last glacial maximum*, ed. H. E. Wright, Jr., J. E. Kutzbach, T. Webb III, E. F. Ruddiman, F. A. Street-Perrott, and P. J. Bartlein, 194–220. Minneapolis: University of Minnesota Press.

Rosen, A. M. 2007. *Civilizing climate: Social responses to climate change in the ancient Near East*. New York: AltaMira.

Ryder, M. L. 1993. Sheep and goat husbandry with particular reference to textile fibre and milk production. In *Bulletin on Sumerian Agriculture* 7, ed. J. N. Postgate and M. A. Powell, 9–32. Cambridge: Sumerian Agriculture Group.

Sanlaville, P. 1992. Changements climatiques dans la péninsule arabique durant le Pléistocène supérieur et l'Holocene. *Paléorient* 18: 5–26.

Schwartz, G. M., H. H. Curvers, F. A. Gerritsen, J. A. MacCormack, N. F. Miller, and J. A. Weber. 2000. Excavation and survey in the Jabbul Plain, western Syria: The Umm el-Marra Project 1996–1997. *American Journal of Archaeology* 104: 419–62.

Schwartz, G. M., and N. F. Miller. 2007. The "crisis" of the late third millennium B.C.: Ecofactual and artifactual evidence from Umm El-Marra and the Jabbul Plain. In *Sociétés humaines et changement climatique à la fin du troisième millénaire: Une crise a-t-elle eu lieu en Haute Mésopotamie?*, ed. C. Kuzucuğlu and C. Marro, 179–203. Varia Anatolica XIX. Paris: De Boccard Édition-Diffusion.

Staubwasser, M., F. Sirocko, P. Grootes, and M. Segl. 2003. Climate change at the 4.2 ka BP termination of the Indus valley civilization and Holocene south Asian monsoon variability. *Geophysical Research Letters* 30: 1425.

Staubwasser, M., and H. Weiss. 2006. Holocene climate and cultural evolution in late prehistoric–early historic West Asia. *Quaternary Research* 66: 372–87.

Steinkeller, P. 1991. The administrative and economic organization of the Ur III state: The core and the periphery. In *The organization of power*, ed. M. Gibson and R. D. Biggs, 15–33. 2nd ed. Chicago: Oriental Institute.

———. 1995. Sheep and goat terminology in Ur III sources from Drehem. In *Bulletin on Sumerian Agriculture* 8, ed. J. N. Postgate and M. A. Powell, 49–70. Cambridge: Sumerian Agriculture Group.

Thomson, E. F. and F. Bahhady. 1995. A model-farm approach to research on crop-livestock integration — I. Conceptual framework and methods. *Agricultural Systems* 49: 1–16.

Twain, M. 1883. *Life on the Mississippi.* Boston: J. R. Osgood and Co.

Van Loon, M. N., ed. 2001. *Selenkahiye: final report on the University of Chicago and University of Amsterdam excavations in the Tabqa reservoir, Northern Syria, 1967–1975.* Uitgaven van het Nederlands Historisch-Archaeologisch Instituut te Istanbul 91. Leiden: Nederlands Instituut voor het Nabije Oosten.

Wallén, C. C. 1968. Agroclimatological studies in the Levant. In *Agroclimatological Methods, Proceedings of the Reading Symposium*, 225–32. Natural Resources Research 7. Paris: UNESCO.

Weber, J. A. 1997. Faunal remains from Tell es-Sweyhat and Tell Hajji Ibrahim. In *Subsistence and settlement in a marginal environment: Tell es-Sweyhat, 1989–1995 preliminary report*, ed. R. L. Zettler et al., 133–67. MASCA Research Papers in Science and Archaeology 14. Philadelphia: Museum Applied Science Center for Archaeology.

Weiss, H. 2000. Beyond the Younger Dryas: Collapse as adaptation to abrupt climate change in ancient West Asia and the eastern Mediterranean. In *Confronting natural disaster: Engaging the past to understand the future*, ed. G. Bawden and R. Reycraft, 75–98. Albuquerque: University of New Mexico Press.

————. 2003. Ninevite periods and processes. In *The origins of Northern Mesopotamian civilization,* ed. E. Rova and H. Weiss, 593–624. Subartu IX. Turnhout: Brepols.

Weiss, H., M.-A. Courty, W. Wetterstrom, F. Guichard, L. Senior, R. Meadow, and A. Curnow. 1993. The genesis and collapse of third millennium North Mesopotamian civilizations. *Science* 261: 995–1004.

Weiss, H., and R. S. Bradley. 2001. What drives societal collapse? *Science* 291: 609–10.

Wilkinson, T. J. 1994. The structure and dynamics of dry-farming states in Upper Mesopotamia. *Current Anthropology* 35 (5): 483–520.

————. 1997. Environmental fluctuations, agricultural production and collapse: A view from Bronze Age Upper Mesopotamia. In *Third millennium BC climate change and Old World collapse,* ed. H. Nüzhet Dalfes, George Kukla, and Harvey Weiss, 67–106. NATO ASI Series I, no. 49. New York: Springer.

————. 2004. *Excavations at Tell es-Sweyhat, Syria 1. On the margins of the Euphrates: Settlement and land use at Tell es-Sweyhat and in the upper Lake Assad area, Syria.* Oriental Institute Publication 124. Chicago: The Oriental Institute of the University of Chicago.

Zettler, R. L. 2003. Reconstructing the world of ancient Mesopotamia: divided beginnings and holistic history. *Journal of the Economic and Social History of the Orient* 46 (1) : 3–45.

Zettler, R. L., J. A. Armstrong, A. Bell, M. Braithwaite, M. D. Danti, N. F. Miller, P. N. Peregrine, and J. A. Weber. 1997. *Subsistence and settlement in a marginal environment: Tell es-Sweyhat, 1989–1995 preliminary report.* MASCA Research Papers in Science and Archaeology 14. Philadelphia: Museum Applied Science Center for Archaeology.

PART III
The Human Element

9

Global Change: Mapping Culture onto Climate

VERNON L. SCARBOROUGH AND WILLIAM R. BURNSIDE

The past is our only barometer of the future, an adage as true for societal development as it is for climate change. But beyond the platitudes, can the past really provide us with effective strategies for altering our habits, behaviors, and mores; or do we continue with our present societal structures and institutional commitments, investing only in the success of our technological creativity and know-how? Can we maintain business as usual, anticipating another set of innovative buffering mechanisms that allow us to further distance ourselves from our biophysical environs and thus continue our overly consumptive ways? Are we structurally doomed to a lock-stepped trajectory that prevents us from changing our exploitative approaches to the environment and one another with the unrealistic hope that we will eventually achieve a better world? Our tunnel vision in repeating our past may result from aligning our assessments with only one history, one emphazing an "empty Earth" myth or a worldview in which natural resources are effectively infinite and human populations never severe enough to truly overexploit at a global level (Beddoe et al. 2009). If resources do thin in one region, there is always access to more elsewhere. Indeed, now our extraction projects are aimed at the moon, Mars, or beyond. But are there limitations (cf. Day et al. 2009)? Is this a worldview that can promote a humanity of equity, happiness, and well-being (cf. Karlsson et al. 2009)? Perhaps there are other pathways that have flourished in the past or exist today that might provide a more nuanced view of what is practical and possible for sustaining our planet (Costanza, Graumlich, and Steffen, eds. 2007).

With greater regularity and focus, archaeologists and researchers from complementary disciplines are gathering to discuss and assess the degree to which climate crises have plagued humanity (Redman et al. 2007). The models and interpretations of the past principally explore correlations between periods of deleterious climate change and responses by human societies, the latter identified by initial levels of catastrophe followed by human adjustment and adaptation—until the next round of external forcings (Bawden and Reycraft, eds. 2000; cf. Scarborough 2000). But all is not doom and gloom, as

studies suggest a "climatic balance sheet" at work, with periods of excessive climatic "pain" offset elsewhere by climatic amelioration and/or economic "gain" (Fagan 2004, 2008; Hodges, Chapter 4). For the present, our evaluative difficulties are twofold: (1) how precise can we become with our dating techniques in "proving" that climate causes practical responses in changing human livelihoods, and (2) is this the question that should actually consume us? Because cultural action and response is as complex and nuanced by its history and preconditioned environments as the near- and long-term climate rollercoaster, can we realistically hope to find our evolutionary trajectory so immediately wedded to a climatic shift? We are all in agreement that climate has been and is a key factor in the fits and starts that have brought our species to the twenty-first century. But other external forces of our own creation, like warfare and overpopulation, pose the same kinds of causal wrangling, i.e., why do some societies "weather" these perturbations and others fragment or collapse?

Perhaps as much time is needed in dissecting the varied and multifaceted aspects of culture and its societal diversity as we presently invest in the nuanced and frequently quantifiable dimensions of climate change. Our penchant to "prove" our predictions (retrodictions in the case of archaeology and history) by way of highly abstract quantitative measures is a testament to our intellectual rigor and creative enterprise—a pathway, more generally, that has provided our species with marvelous breakthroughs in longevity, health, and degrees of well-being for many segments of the human population globally, and especially in the very recent past. This similar approach to climate-change analysis is a meaningful and significant orientation. Nevertheless, it has its limitations.

We suggest that several pathways in the ancient past have directed the manner by which we use and concentrate our many resources, and an assessment of one of these pathways that is different from our own is worthy of discussion here. If evolutionary theory has taught us anything, it is that diversity provides the biophysical wherewithal to accommodate adaptation in the face of unremitting change. What might be another societal strategy for longevity and success? Are there aspects of such an adaptation that might aid us in coping better with our own present and future?

The ancient Maya and living Balinese have, or once did have another trajectory unlike our own, one that has been nearly completely co-opted or appropriated by the current nation state (Scarborough 2006, 2008) (Fig. 1). That trajectory represented a highly complex socioeconomic and sociopolitical suite of adaptations to a semitropical environment. Although a world apart both geographically and temporally, these two case studies suggest that such a pathway to social complexity was neither aberrant nor unique—just highly diminished today.

In both cases, the biophysical environs provide the initial conditions that stimulated aspects of complexity (Lansing 2003; Scarborough and Burnside 2010). Tropical settings are renowned for their abundance of life forms and the species' diversity manifested (Fig. 2). Nevertheless, within any one geographical patch or microenvironment few representatives of a particular

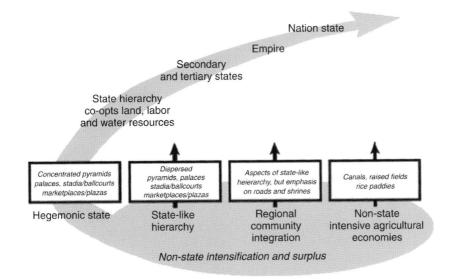

Nation state

Empire

Secondary
and tertiary states

State hierarchy
co-opts land, labor
and water resources

| Concentrated pyramids palaces, stadia/ballcourts marketplaces/plazas | Dispersed pyramids, palaces stadia/ballcourts marketplaces/plazas | Aspects of state-like heierarchy, but emphasis on roads and shrines | Canals, raised fields rice paddies |

Hegemonic state

State-like
hierarchy

Regional
community
integration

Non-state
intensive agricultural
economies

Non-state intensification and surplus

Landscape Signatures

Figure 1. Landscape signatures. Kinds of engineered landscapes and associated socioeconomic and sociopolitical organizational schemes appropriated by hegemonic states with time.

species are represented, leading to a highly dispersed distribution of species members across a set of environments. These relatively low population densities—or the actual number of plants or animals of a given species within a patch of a given size—forces dispersion of organic resources and impacts a socially dependent species like humans (Scarborough 2000).

One of the principal hallmarks of conventionally assessed civilization and social complexity is our ability to capture and collect raw and subsequently highly refined resources, and then to concentrate them for redistribution, and, in today's world, to market them. The notion of urbanism as first established along the lower Tigris and Euphrates by at least 3500 B.C. reveals the success of this trajectory—one that continues to underpin our current orientation or worldview of an "empty Earth" (though the Earth is no longer empty). The ability of Sumerians to use the technological breakthroughs of the wheel, the sail, metallurgy, and canalization collectively to exploit the landscape in ways never before possible ushered in an "urban revolution" as V. Gordon Childe (1950) called it. Irrigation agriculture as a function of canalization was perhaps the most profound breakthrough in that "marginalized wasteland" converting it to "a desert abloom." The context for an empty Earth was established, and we have not backed away from this trajectory since. The legacy of salinization or deleterious sedimentation rates as a consequence of river diversion and canalization reveals the long-term costs frequently unanticipated with technological breakthroughs. Although resources are rapidly

exploited and consumed, entire regions can be denuded and their occupants required to relocate. This pattern has been modified by today's nation state in which resource exploitation occurs away from the largest and most developed states, while the mining of resources and landscapes goes unabated in economically marginalized zones associated with undeveloped institutional structures but containing abundant natural resources—both in terms of materials and labor.

So how can a tropical environment inform the discussion? In both contemporary Bali and the ancient southern Maya Lowlands—the latter dating to a period from 600 B.C. to A.D. 900 for our purposes—society settled initially into its biophysical ecology by incrementally modifying its surrounds (Scarborough 2008; cf. Scarborough 1993) (Figs. 3 and 4). Unlike the rapid and frequently dramatic set of changes associated with technological and landscape alterations, the early sedentists in these semitropical settings were constrained by the immediate environmental limitation for making a living. Because of the dispersed availability of resources, humans were required to disperse as well. The first towns and "cities" in the Maya area were not nucleated centers of wealth concentration similar to the first cities of the Near East or elsewhere along the great drainages of semiarid Asia inclusive of the Nile, the latter again revealing the impressive role of formalized irrigation. What we find in both Bali and the Maya Lowlands is a huge population base, but one spread out widely across the forested landscape in a manner mimicking the tropical rhythms and tempo. The largest cities were likely a full magnitude less populous than the earliest cities of the Near East, but the semitropical hinterlands were likely a magnitude more densely populated than those rural settings also identifiable in the Near East (Scarborough 2003a, 2005, 2006). But given our definition of resource concentrations as the linchpin for social complexity, how is it that these semitropical societies not only survived, but flourished?

The answer rests in their ability to rapidly adapt to the constantly changing ecology of heaving/breathing sets of abundant life forms composing a tropical setting. The flexible networks connecting humans to their environs and to one another allowed a high degree of interdependency (Scarborough and Valdez 2009). This was manifest by way of scheduling and rapid movement of goods, services, and information. Because of high temperatures and humidity, the storage of organic matter is inhibited in a tropical regime. Although measures evolved to promote some degree of storage—resource concentration—by investing in certain root crops like manioc that can be left in the ground for up to three years, or extracting vast supplies of salt for preserving fish and related foodstuff, the most efficient adaptations involved timing and reducing distances (Scarborough 2008). Much has been made of the elaborate calendrical systems of both the Maya and the Balinese in terms of complex ritualized assessments of the heavens and the role that both deities and kings played in promoting prediction in controlling their courts and support populations. Nevertheless, a more mundane origin for this calendar is posited as linking the necessity of scheduled economic activities and the movement of perishable goods (Scarborough 2008). Too, we are now beginning to understand the extensiveness of the road system in the

Figure 2. Semitropical setting near Tikal, Guatemala—the ancient Maya heartland.

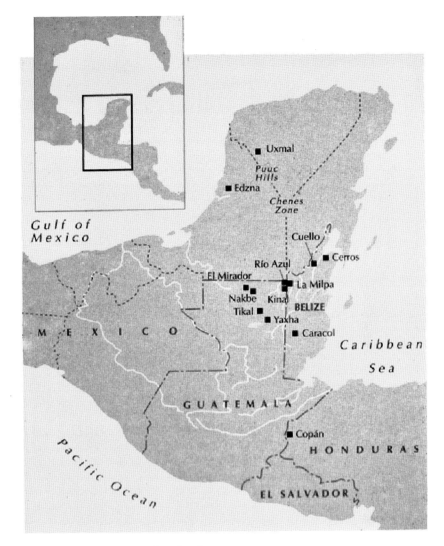

Figure 3. Map of the Maya Lowlands.

Maya Lowlands, and its implications for shortening the time human bearers might carry the scheduled arrival of perishables for rapid consumption (Shaw 2001, 2008). Because Bali was a small island—150 km E/W and 110 km N/S— the movement of resources was less dependent on formalized roads (Fig. 4).

What becomes clear is that a different set of interactions evolves in semi-tropical settings that inform how society will develop. Because of the relatively decentralized character of settlement and the difficulties with storage, society intensifies its harvesting and consumption of resources by producing and transporting them rapidly. These social and biophysical relationships are

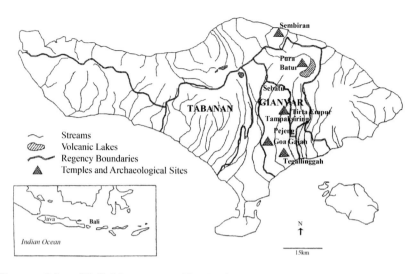

Figure 4. Map of Bali. Map prepared by Andras Nagy.

heterarchically identifiable (Crumley 1995) by a much less steep-sided hierarchical order than our own, with highly malleable linkages between points of supply and demand (Scarborough and Valdez 2003). In a word, they are resilient and capable of adjusting situationally without fragmentation. The ability of a society to self-organize onto its changing ecology, an ecology altered significantly by society's own changing demands, produces a "loose-knit, glove fit" association within and between groups and the landscapes they inhabit (perhaps in contradiction to our attachment to the "invisible hand" metaphor). An accretional series of changes slowly modifies the landscape to best accommodate a society, but society takes on the understanding that the environs are an indivisible signature of itself (Fig. 5). In Bali, the religion most practised is that dedicated to "agama tirta" or "holy water" as the defining element in its complex rice paddy harvests (Scarborough, Schoenfelder, and Lansing 1999).

In both Bali and the Maya Lowlands, labor is the key organizing principle. Where technological breakthroughs stimulate major periods of societal change in our worldview, labor organization and careful task preparation dictate the trajectories of the semitropical societal examples. By maintaining skill-oriented, group interdependent activities on an engineered landscape over generations, an environment and its people incrementally develop/change. In earlier assessments of complex societal or institutional "systems," hypercoherence was viewed as a highly deleterious outcome resulting in internal societal collapse (Flannery 1972). For those societies evolving in tropical settings, however, the flexibility to incorporate another information source or useful method/approach to better harvest the environs was promoted. Those ideas that did not work in the specific interrelationships to a microenvironment were

Figure 5. Balinese landscape.

quickly abandoned, while those of productive ends were further cultivated or developed. In Bali, this resulted in the highly intricate, self-organized rice paddies and their dizzily climbing stair-stepped surfaces on an otherwise rugged topography, but it was an accretional enterprise of at least 400 years, likely initiated by the eleventh century (Scarborough 2008; Scarborough, Schoenfelder, and Lansing 1999). And it continues today (Lansing 2006) (Fig. 5).

To graphically display these different trajectories to social complexity, let us examine Figure 6. The upper diagram emphasizes the nonlinear escalation of both material and social costs to a society with increasing concentrations of power and levels of hierarchical control—leading precipitously to greater social complexity. This model shows the effects of technological breakthroughs that lift complexity, but at extremely high costs, sometimes resulting in significant fragmentation of society and reversion to lesser conditions. This is conventional wisdom charted by most versions of Western history stretching back to the early city-states along the Tigris-Euphrates. The lower display attempts to capture the more incremental, long-lasting and monitored assessment of the engineered landscape by way of an active labor pool with skill sets promoted generationally and only evolving in refinement to local conditions. It reveals a less "pitched" trajectory to social complexity based on less dramatic production and consumption growth than the former graph. It best mimics the interdependencies embedded in tropical ecosystems and the networks accommodating self organization.

If tropical settings for the development of complex society were so resilient, why are so few in existence, and why did the archetypical semitropical civilization—the ancient Maya—so completely collapse? At the outset, the nation-state is highly hegemonic and coercive, making most other pathways to complexity vulnerable (see Fig. 1). The biophysical fragility of a tropical environment is not one in which our nation-state model wishes to invest nuanced understanding—just reflect on the clear-cutting of tropical forests today for ungulates first domesticated in the Near East. Exploitation and recent colonization efforts are no match for an internally self-organized society less wedded to technological breakthroughs (Scarborough 2006).

In the Maya case, however, we have internal collapse without the interjection of a Western nation-state—500 years before the Spanish colonial embrace. What went so wrong? Although the Maya had a longevity of at least 1500 years, their collapse was significant and very real. To emphasize the dramatic changes for this otherwise highly resilient society, only 500,000 people occupy the southern Maya Lowlands today—a region of perhaps 100,000 km^2 that is estimated to have had a population of 5 million at A.D. 700 (Scarborough and Burnside 2010—after Rice and Culbert 1990). Recognizing the scope of the calamity reveals a great deal about the highly altered built environs and the skill-oriented labor that self organized into this setting (Scarborough 2003b).

Like Bali's volcanic setting today, the Maya slowly, incrementally, altered their karstic terrain. Although we far from understand the extent of landscape change, it is clear that the 40 percent of the Maya Lowlands that is currently unusable seasonal swampland was at least partially reclaimed (Dunning and Beach 2000). Like Bali, when rice paddies go unclaimed and unkempt,

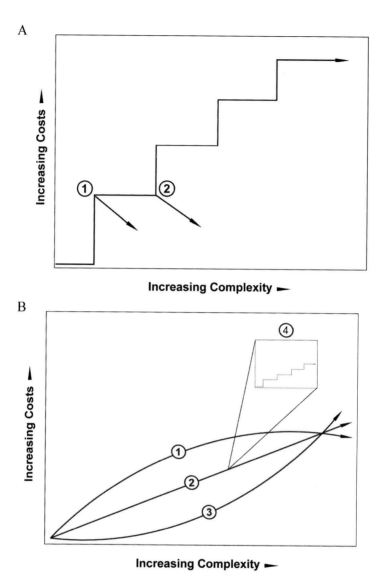

A

Increasing Costs ↑

Increasing Complexity ▬

B

Increasing Costs ↑

Increasing Complexity ▬

Figure 6. Schematic graphs identifying two idealized types of complexity trajectories. A. A high-stepped, steeply pitched pathway frequently associated with technological investment. A1 represents the juncture of a successful phase transition to a new level of social complexity, though vulnerability exists. A2 represents the onset of change, either a phase transition to new levels of complexity or a partial reversion to an earlier adaptation or possible collapse. B. A highly monitored, low-pitch, self-organizing pathway frequently associated with labor investment. B1, B2, and B3 represent possible complexity trajectories based on initial conditions and environmental perturbations. B4 reveals the same stepwise forces advancing change, but operating at an increased frequency and a much reduced ascent.

rapid slope erosion results in the burial of ancient swamp-margin fields and the loss of upslope terraces or bermed downslope fields (Deevey et al. 1979; Dunning et al. 1999, 2002, 2003). The generations of preparation by the Classic period Maya in erecting the intricacies of a highly complex, interdependent and productive human ecology was disrupted abruptly. The subsistence and economic system was predicated on the generations of self-organization into the landscape. When information was not coordinated and schedules were misunderstood, the otherwise resilient set of adaptations became highly vulnerable (cf. Scarborough 2003a, 2003b).

In the Balinese example, farmers today have self organized into rice co-operatives or *subak* units (Lansing 1991, 2006). The cooperatives derive their support from several villages. Each village spreads its risk by having its members dispersed and loyal to more than one subak, though each villager is committed to only one subak. During festivals and ritualized gatherings at the numerous water temples across the island, subak members consult with priests and related functionaries about ancestors and gods, but also exchange information about the conditions of fields. Functionaries review not only auguries and calendars but also the information conveyed by each supplicant concerning the rice harvest. Acting not unlike agricultural extension officers and repositories for practical local knowledge, these functionaries provide information about the limitations of water availability or the likelihood that a specific farmer's field will be infested by a neighbor's blight (Lansing and Kremer 1993). Planting and harvesting rules are meted out and schedules put in place. In other words, water temples and the personages coordinating them in this ritualized economy are heavily invested in the direction and movement of quality information. This concentration and refinement of redistributed information underscores the kinds of interdependencies that have evolved. The social and physical distances between core and periphery, urban and rural, even "haves" and "have nots," is less pronounced than in our early nation-state trajectory (Scarborough 2005; Scarborough, Schoenfelder, and Lansing 1999).

For the ancient Maya, the hubris of the centers and the kings that coordinated those "cities" seems to have grown by the eighth century (Martin and Grube 2000). Clearly, the population was larger than it had ever been before, and the demands on the engineered landscape were considerable. When coupled with the set of droughts that struck the Maya region from AD 760 to 910 (Haug et al. 2003), a perfect storm was unleashed. Although the Maya had been extremely adept and clever in cultivating their jungle ecology through centuries of self organization, it was no match for what was to unfold. And when the coordinating role of the priest-kings and nobility closed themselves to their interdependency role in listening to the voices of their hinterlands (Scarborough 2007)—the sustaining population of skilled and highly adept laborers—the terrace walls collapsed and the richly productive and reclaimed wetlands were buried under tons of sediment and stone (Curtis et al. 1998; Deevey et al. 1979; Dunning and Beach 2000; cf. Anselmetti et al. 2007; Webster et al. 2007; Yaeger and Hodell 2008). Without the highly scheduled workforce maintained by the open, though structured, exchange of landscape

information from the coordinating nodes—like the water temples in Bali—the great centers were abandoned and the hinterlands left further to the vagaries of tropical degradation (Scarborough 2003a, 2003b).

Returning to the climate crises in the human past, how does this brief presentation address our concerns about the present—or perhaps the future?

(1) Climate change will move geographic distributions of important species and suites of species, including those on which we depend, such as agricultural crops, as well as those that we contain, like obligate human parasites. Nevertheless, current approaches that focus on modeling climate change scenarios using "niche modeling" to assess how changing climate will affect distributions—from cereal production to malaria contraction—do not evaluate sociocultural responses to these volatile distributions. Consensus climate models are too conservative because species often adjust in idiosyncratic ways, frequently conditioned by human behavior with novel species' combinations arguably meriting local, ecologically informed evaluations.

(2) Decades of research in archaeology and ecology—from the collapse of the Maya and the fall of the Roman Empire to the shortcomings of the Green Revolution—point to both the self-destructive and erosive power of societies as well as the wisdom of adapting sociocultural responses to local conditions.

(3) Shifting climates and associated resource regimes will ignore existing political boundaries, perhaps causing mass famines and human migrations, the latter meriting an immediate cultural-ecological perspective.

The principal point is that the historic and even prehistoric organizational decision-making of a society is significantly conditioned by the landscape it occupies and the methods that society chooses to harvest or exploit its resources. This complicated set of interdependencies is as telling in assessing the effects of any external forcing on that society. Degrees of societal resiliency or vulnerability to climate change are affected by a suite of variables embedded in the economic, political, and ideological spheres of a culture, but how a group defines, parses, and then weighs these variables becomes the difference between survival and collapse. One of the most long-lived and resilient societies on record was the ancient Maya. Though their culture continues, its former landscape signature collapsed in a manner of near completeness. The lesson here is that no cultural pathway is impervious to catastrophe; but truly learning from the past might well ameliorate the effects of our climate crises.

References

Anselmetti, F. S., D. A. Hodell, D. Ariztegui, M. Brenner, and M. F. Rosemeier. 2007. Quantification of soil erosion rates related to ancient Maya deforestation. *Geology* 35: 915–18.

Bawden, G., and R. Reycraft, eds. 2000. *Natural disaster and the archaeology of human response*. Albuquerque: Maxwell Museum of Anthropology and the University of New Mexico Press.

Beddoe, R., R. Costanza, J. Farley, E. Garza, J. Kent, I. Kubiszewski, L. Martinez, T. McCowen, K. Murphy, N. Myers, Z. Ogden, K. Stapleton, and J. Woodward. 2009. Overcoming systemic roadblocks to sustainability: the evolutionary redesign of worldviews, institutions, and technologies. *Proceedings of the National Academy of Sciences* 106: 2483–89.

Childe, V. G. 1950. The urban revolution. *Town Planning Review* 21: 3–17.

Costanza, R., L. J. Graumlich, and W. Steffen, eds. 2007. *Sustainability or collapse? Integrated history and future of people on earth (IHOPE)*. Dahlem Workshop Report 96R. Cambridge, Mass.: The MIT Press.

Crumley, C. L. 1995. Heterarchy and the analysis of complex societies. In *Heterarchy and the analysis of complex societies*, ed. R. M. Ehrenreich, C. L. Crumley, and J. E. Levy, 1–6. No. 6 of *Archaeological Papers of the American Anthropological Association*. Arlington, Va.

Curtis, J., M. Brenner, D. A. Hodell, R. A. Balser, G. A. Islebe, and H. Hooghiemstra. 1998. A multi-proxy study of Holocene environmental change in the Maya Lowlands of Peten, Guatemala. *Journal of Paleolimnology* 19: 139–59.

Day, J. W., Jr., C. A. Hall, A. Yanez-Arancibia, D. Pimentel, C. Ibanez Marti, and W. J. Mitsch. 2009. Ecology in times of scarcity. *Bioscience* 59 (4): 321–31.

Deevey, E. S., D. S. Rice, P. M. Rice, H. H. Vaughan, M. Brenner, and M. S. Flannery. 1979. Maya urbanism: impact on a tropical Karst environment. *Science* 206: 298–306.

Dunning, N., and T. Beach. 2000. Stability and instability in prehispanic Maya landscapes. In *An imperfect balance: Precolumbian new world ecosystems*, ed. D. Lentz, 179–202. New York: Columbia University Press.

Dunning, N., J. G. Jones, T. Beach, and S. Luzzadder-Beach. 2003. Physiography, habitats, and landscapes of the Three River region. In *Heterarchy, political economy, and the Ancient Maya*, ed. V. L. Scarborough, F. Valdez Jr., and N. Dunning, 14–24. Tucson: University of Arizona Press.

Dunning, N., S. Luzzadder-Beach, J. Jones, V. L. Scarborough, and T. P. Culbert. 2002. Arising from the Bajos: anthropogenic change in wetlands and the rise of Maya civilization. *Annals of the Association of American Geographers* 92: 267–83.

Dunning, N., V. L. Scarborough, F. Valdez Jr., S. Luzzadder-Beach, T. Beach, and J. G. Jones. 1999. Temple mountains, sacred lakes, and fertile fields: ancient Maya landscapes in northwestern Belize. *Antiquity* 73: 650–60.

Fagan, Brian. 2004. *The long summer: how climate changed civilization*. New York: Basic Books.

———. 2008. *The great warming: climate change and the rise and fall of civilizations*. New York: Bloomsbury.

Flannery, K. V. 1972. The cultural evolution of civilizations. *Annual Review of Ecology and Systematics* 3: 399–426.

Haug, G. H., D. Gunther, L. C. Peterson, D. M. Sigman, K. A. Hughen, and B. Aeschlimann. 2003. Climate and the collapse of Maya civilization. *Science* 299: 1731–35

Karlsson, S. I., M. Cesario, S. Anthony, K. Arabena, E. Assadourian, S. Ceausu, R. Constanza, J. Mohammed-Katerere, W. Pan, V. Scarborough, S. Sing,

J. Timm, and U. de Zoysa. 2009. Group report: towards a world of well-being: shared goals and diverse pathways. In *The Planet in 2050*, ed. W. Steffen, G. Brasseur, T. Jacobsson, and R. Costanza. In Press.

Lansing, S. J. 1991. *Priests and programmers: technologies of power in the engineered landscape of Bali*. Princeton: Princeton University Press.

———. 2003. Complex adaptive systems. *Annual Reviews in Anthropology* 32: 183–204.

———. 2006. *Perfect order: recognizing complexity in Bali*. Princeton: University of Princeton Press.

Lansing, J. S., and J. N. Kremer. 1993. Emergent properties of Balinese water temples. *American Anthropologist* 95: 97–114.

Martin, S., and N. Grube. 2000. *Chronicle of Maya kings and queens*. New York: Thames and Hudson.

Redman, C. L., C. L. Crumley, F. Hassan, K. F. Hole, F. Riedel, J. A. Tainter, P. Turchin, and Y. Yasuda. 2007. Group report: millennial perspectives on the dynamic interaction of climate, people, and resources. In *Sustainability or collapse? Integrated history and future of people on earth (IHOPE)*, ed. R. Costanza, L. Graumlich, and W. Steffen, 115–48. Dahlem Workshop Report 96R, Cambridge, Mass.: The MIT Press.

Rice, D. S., and T. P. Culbert. 1990. Historical contexts for population reconstruction in the Maya Lowlands. In *Precolumbian population history in the Maya Lowlands*, ed. T. P. Culbert and D. S. Rice, 1–36. Albuquerque: University of New Mexico Press.

Scarborough, V. L. 1993. Water management in the Southern Maya Lowlands: an accretional model for the engineered landscape. In *Economic aspects of water management in the prehispanic new world* (Research In Economic Anthropology, Supplement 7), ed. V. L. Scarborough and B. L. Isaac, 17–69. Greenwich, Conn.: JAI Press.

———. 2000. Resilience, resource use, and socioeconomic organization: a Mesoamerican pathway. In *Natural disaster and the archaeology of human response*, ed. G. Bawden and R. Reycraft, 195–212. Albuquerque: Maxwell Museum of Anthropology and the University of New Mexico Press.

———. 2003a. *The flow of power: ancient water systems and landscapes*. Santa Fe: School of American Research Press.

———. 2003b. How to interpret an ancient landscape. *Proceedings of the National Academy of Sciences* 100: 4366–68.

———. 2005. Landscapes of power. In *A catalyst for ideas: anthropological archaeology and the legacy of Douglas W. Schwartz*, ed. V. L. Scarborough, 209–28. Santa Fe: School of American Research Press.

———. 2006. Intensification and the political economy: a contextual overview. In *Agricultural strategies*, ed. J. Marcus and C. Stanish, 401–18. Los Angeles: Cotsen Institute of Archaeology at UCLA.

———. 2007. The rise and fall of the ancient Maya: a case study in political ecology. In *Sustainability or collapse? Integrated history and future of people on earth (IHOPE)*, ed. R. Costanza, L. Gramlich, and W. Steffen, 51–60. Dahlem Workshop Report 96R. Cambridge, Mass.: The MIT Press.

———. 2008. Rate and process of societal change in semi-tropical settings: The ancient Maya and the living Balinese. *Quaternary International* 184: 24–40.

Scarborough, V. L., and W. R. Burnside. 2010. Complexity and sustainability: perspectives from the ancient Maya and the modern Balinese. *American Antiquity* 75: 327–63.

Scarborough, V. L., J. W. Schoenfelder, and J. S. Lansing. 1999. Early statecraft on Bali: the water temple complex and the decentralization of the political economy. *Research in Economic Anthropology* 20: 299–330.

Scarborough, V. L., and F. Valdez Jr. 2003. The engineered environment and political economy of the Three Rivers region. In *Heterarchy, political economy, and the ancient Maya: the Three Rivers region of the east- central Yucatan Peninsula*, ed. V. L. Scarborough et al., 3–13. Tucson: University of Arizona Press.

———. 2009. An alternative order: The dualistic economies of the ancient Maya. *Latin America Antiquity* 20 (1): 207–27.

Shaw, J. M. 2001. Maya Sacbeob: form and function. *Ancient Mesoamerica* 12: 261–72.

———. 2008. *White roads of the Yucatan.* Tucson: University of Arizona Press.

Yaeger, J., and D. A. Hodell. 2008. The collapse of Maya civilization: assessing the interaction of culture, climate, and the environment. In *El Nino, catastrophism, and culture change in ancient America*, ed. D. H. Sandweiss and J. Quilter, 187–242. Cambridge, Mass.: Harvard University Press.

10

Climate, Crisis, Collapse, and Ancient Maya Civilization: An Enduring Debate

NORMAN HAMMOND

The abandonment of the Maya cities of the Central American rain forest has been a matter of interest and attempted explanation since the Maya were first brought to wide public attention by John Lloyd Stephens in 1841. Stephens speaks (1841, 1:158–59) of the monuments of the great site of Copan "standing in the depth of a tropical forest, silent and solemn, strange in design, excellent in sculpture, rich in ornament, different from the works of other people . . . their whole history so entirely unknown, with hieroglyphs explaining all, but perfectly unintelligible. . . . In regard to the age of this desolate city, I shall not at present offer any conjecture. . . . Nor shall I at this moment offer any conjecture in regard to the people who built it, or to the time when or the means by which it was depopulated, and became a desolation and ruin; whether it fell by the sword, or famine, or pestilence. The trees which shroud it may have sprung from the blood of its slaughtered inhabitants; they may have perished howling with hunger; or pestilence, like the cholera, may have piled its streets with dead, and driven forever the feeble remnants from their homes; of which dire calamities to other cities we have authentic accounts. . . . One thing I believe, that its history is graven on its monuments."

More than a century and a half later, the debate is still a lively one with no sign of final resolution. This notorious "collapse" of the Classic Maya in the ninth and tenth centuries has remained a perennial subject of interest (reviewed most recently by Aimers 2007, and Yaeger and Hodell 2008). Some scholars (notably in Demarest et al. 2004) have attempted to substitute a terminology of "transformation" for the more dramatic "collapse" to mark the fact that substantial, although less urban, populations remained in parts of the southern lowlands, and that the polities of Yucatan continued to flourish into the early second millennium AD and were still functioning in some form at the time of the Spanish conquest, albeit from different capitals and with a less complex political structure.

The great cities of the tropical forest were, whatever terminology is applied, abandoned in the ninth century. Construction of public buildings including palaces and temples, patronage art on both the communal and personal scale,

and dedication of dated monuments bearing the images and deeds of divine kings all ceased; few cities have Long Count dates after 10.3.0.0.0. (AD 889), none after 10.5.0.0.0 (AD 928) (for an explanation of the Maya calendar, see Sharer 2006: 102–20).

Possible reasons for what has become known as the "Classic Maya Collapse" have ranged from the all-embracing and environmental to the specific and social. The Collapse marks the end of the Classic Period of AD 300–900, during which the civilization attained its apogee and cities such as Copan, Tikal, Palenque, and Calakmul flourished in the tropical lowlands of southern Mexico, Guatemala, Belize, and westernmost Honduras and El Salvador (see Fig. 1). The debate about the Collapse has centered on what are known as the Southern Lowlands, with both the Highland region to the south and the Yucatan Peninsula to the north playing a decidedly minor role, and has focused on the period from AD 750 to 1050.

Explanations advanced over the past century and a half—from Stephens' "sword, famine or pestilence" onward—have ranged from the political (invasion, insurrection, internecine warfare, and social decay) to the natural (including soil exhaustion and crop failure, plant and human disease, and sudden catastrophes, including hurricanes and earthquakes [Culbert 1973]). Most of these monocausal explanations have been discarded from lack of evidence or, increasingly, from evidence that simple answers are inadequate. Combination theories that have been put forward include larger and more closely packed cities overstretching both the productive capacity of the landscape and the managerial capacity of the Maya political structure to deliver enough sustenance to increasingly stressed urban populations (Hammond 1982: 140–41). In such complex multicausal models, the potency of the various factors proposed has varied widely from author to author. There has also been some correlation between interpretations of the ancient Maya and broader modern political and environmental concerns (Wilk 1985; see also Sabloff 1990: 166–75 and Diamond 2005).

Environmental explanations for the collapse in the southern lowland rain forest zone were among the earliest explicit hypotheses: Cook (1921) suggested that the traditional slash-and-burn *milpa* agriculture of the historic Maya would have pushed soils toward exhaustion and grass encroachment as populations became larger and fallow periods enforcedly shorter. Cooke (1931) thought that soil erosion was a concomitant of the milpa regime, and that this not only made land uncultivable but also silted up shallow lakes to create the *bajo* swamps found across the Maya lowlands, removing useful water resources from the economic web (although Harrison [1977] later argued for the utility of bajo soils in Maya farming).

A landmark in the discussion came in 1970, with an advanced seminar held at the School of American Research in Santa Fe, N.M.; the resulting volume codified the outcome (Culbert 1973; Willey and Shimkin 1973). In this discussion, Adams (1973, 22) laid out the observed archaeological correlates, summarized in terms of "the failure of elite-class culture": the cessation of architectural construction, inscribed and dated monuments, and sumptuary art together with an apparent rapid depopulation of both town and countryside,

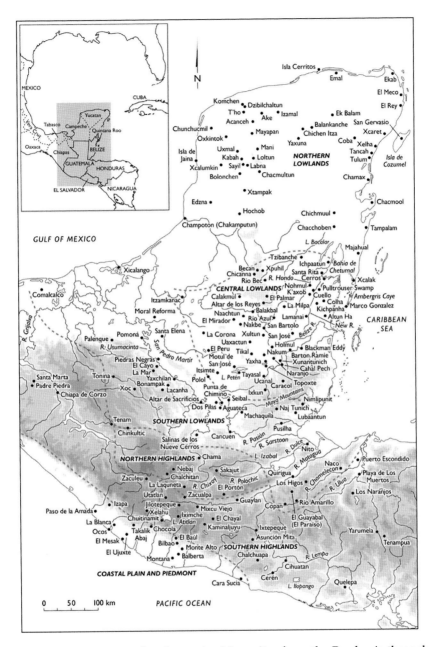

Figure 1. Location map showing major Maya sites from the Preclassic through Postclassic periods, 1000 BC–AD 1500 (after figure 1.1, *The Ancient Maya*, 6th ed., by Robert J. Sharer and Loa P. Traxler © 2006 Stanford University Press).

all within a period of less than a century. Adams also (1973, 23–33) organized potential causes advanced over the previous half century by a variety of scholars into seven categories:

1. Ecological, including soil exhaustion, erosion, savannization, and water loss;
2. Catastrophic, including earthquake and hurricane;
3. Evolutionary;
4. Disease;
5. Demographic;
6. Social dissolution; and
7. Invasion.

Sabloff (1973, 36) took more or less the same range of causes from the same debate, but reorganized them in terms of a pragmatic criterion of testability. Sabloff's two basic classes were of internal and external causes. Internal causes were divided into "natural" and "sociopolitical," with sociopolitical then subdivided into "peasant revolt" and "intersite warfare." External causes were seen as either "economic" or "sociopolitical," the latter here seen as invasion with or without resettlement. Sabloff's list of natural causes was essentially the same as Adams's, but brought in "insect pests" as a new category, and also introduced "climatic change" for the first time in those specific terms (although Adams's "water loss" clearly intends something similar, and notes an explanation originating with Cooke [1931]).

Almost four decades of further archaeological investigation in the Maya lowlands have whittled down Adams's and Sabloff's lists somewhat: neither hurricanes nor earthquakes, both essentially very short-term afflictions, are taken seriously today; nor have disease (of plants or humans) and insect pests proved viable theses. Of the sociopolitical causes, invasion from outside the Maya area (seriously argued last by Sabloff and Willey [1967], although recently resurrected as the cause of sociopolitical change in the Early Classic period, and hence perhaps due for re-examination) has been eclipsed by demonstrable internecine warfare between Maya kingdoms, as the decipherment of Maya texts has shown the wide extent of strife (Martin and Grube 2000). Any peasant revolts have been subsumed within broader models of social dissolution in which they are consequent, not causal; the evidence advanced for them—the obliteration of the rulers' faces on many stelae (Hamblin and Pitcher 1980)—seems in any case to have been the result of deconsecration rather than willful destruction, as the care given to such destruction, and in some cases the subsequent enshrinement of the mutilated monuments, indicates. Those causes still taken seriously today include demography—a late Classic population explosion outrunning environmental and managerial capacity (Culbert and Rice 1990)—and warfare (Demarest 1997; Webster 2002).

A group of theories that has received renewed attention and some cautious acceptance as at least partial explanation for the Maya Collapse involves climate change, especially severe episodic drought leading to subsistence fail-

ure (Gill 2000). The notion of catastrophic climate change as a salient cause had been mooted earlier, but discounted: Maya historical records had indeed complained of episodic droughts and crop failure, but the Maya had survived to tell the tale. Then in 1995 oxygen-isotope data from cores in Lake Chichankanab in Yucatan suggested a two-century drought from AD 800 to 1000, the period of the collapse (Hodell et al. 1995). This was elaborated by Gill (2000) into a general explanation for the demise of Classic Maya civilization. However, one problem with Gill's explanation is that it does not account for such phenomena as the dramatic florescence of the Puuc cities (such as Uxmal and Sayil) and the initial part of Chichén Itzá's even more striking rise to power in Yucatan, both of which occurred during the two centuries between AD 800 and 1000.

As Yaeger and Hodell (2008) point out, the term "collapse" is perhaps misleading; the Maya demise took well over a century, even on Gill's own criterion of last-dated-monument dedication. In any case, this criterion is unreliable because there is no consistent correlation between the date of the latest (discovered) monument at a site and its abandonment as a community. In addition, there are problems of coarse chronology within the drought period: even the "best" radiocarbon dates, with low error deviations, span a period of a century; and archaeological ceramic chronologies based on stylistic change in the vessels themselves correlated with the stratigraphic contexts in which their remains are found have phases more than a century in length. How can these disparate data be reconciled, if at all? Some recent palaeoclimatological evidence (*infra*) allows us at least an attempt.

The alternating seasons of the Maya year are governed by the migration of the meteorological equator, when the tropical trade winds shift north and south, giving a dry season from January to May and a wet one from June to December, with as much as 90% of the rainfall within a three-month period of that wet season. The effect of this shift has been documented in a 170-m core from the Cariaco basin off Venezuela, where anoxia has left the thin laminae of annual deposits undisturbed by burrowing marine organisms (Haug et al. 2003; Peterson and Haug 2005). Changes in titanium and iron levels stand proxy for the amount of seasonal runoff—effectively an ancient rain gauge. The Cariaco core showed a broad band of low rainfall over the AD 800–1000 period, but also four short periods of intense drought within that, each lasting five to six years and separated by around half a century; the fine detail of that reconstruction has allowed the coarser resolution of Maya Area sediment cores to be more subtly interpreted.

During these short intense droughts, groundwater sources would have been vital for community survival, since reservoirs could not have been replenished. In the wetter southern part of the Maya lowlands, where the great cities were abandoned earliest, the water table lies deeply buried. Only in Yucatan does it come close enough to the surface to be accessible through caves and *cenotes*, which might explain the continued existence (though not the evident prosperity) of the Puuc cities and Chichén Itzá. On the other hand, a quite drastic reduction in rainfall in the southern area would still leave an adequate margin for maize cultivation, while in Yucatan, with its exiguous

rainfall (<500 mm today in the northwestern corner), even a modest reduction might seriously inhibit such vital processes as germination.

Gill's model of collapse (2000; Gill et al. 2007), based on the assumption that the latest dated monuments at Maya sites can stand proxy for their abandonment, fits quite well into Peterson and Haug's more precise schema (2005), with its drought maxima around AD 760, 810, 860, and 910. The first of these dates coincides with the estimated maximum Maya population level in AD 750–800, when more people were crammed into more and larger cities, more closely packed into the landscape than ever before, and competing more ferociously for resources of land and labor. The second through fourth drought maxima occur as the number of monuments dedicated, and the number of cities erecting them, declines with increasing speed after AD 800. Hodell et al. (2005), working with further data from Lake Chichankanab, have refined their thesis of a decade earlier (Hodell et al. 1995) to suggest two major droughts, one from AD 770 to 870 with two half-century cycles, the second from AD 920 to 1100 in four similar cycles, separated by a moister interval. This model is fairly compatible with the more precise Cariaco Basin data. By the time the second cycle began, no Maya cities were still dedicating dated monuments, and few in the southern lowlands show more than minor evidence of continued habitation.

At Lake Punta Laguna, further northeast and in the wetter, more forested zone of Northern Quintana Roo, Hodell et al. (2007) also found evidence for lower lake level and drier climate at the period spanning the collapse (AD 750–1050); but also they found similar phenomena to have occurred at about the same times as earlier major discontinuities in Maya cultural history: the Preclassic abandonment (AD 150–250), when major centers such as El Mirador and Nakbe were permanently abandoned, and others, including Seibal and Nohmul, underwent drastic shrinkage of population; and the "Classic Hiatus" (AD 534–593), a period when few monuments were dedicated, but which has recently been ascribed to widespread warfare.

Perhaps political and environmental causes for the drastic changes in Classic Maya society over the Collapse period cannot be, and should not be, separated. The failure of crops could have been seen as a failure of the divine kingship to protect the people from the malignity of personified nature, which could have resulted in active social unrest—manifested in internal revolt or attacks on neighbors—and passive resistance. People voting with their feet—moving out to find subsistence in the countryside—would rapidly remove the underpinnings of elite urban culture. Construction of temples and dedication of monuments would cease before actual abandonment and, therefore, appear more precipitate. We have found this at the large city of La Milpa in northwestern Belize (population estimated at 50,000 in AD 800), where several major royal construction projects including enlargement of the palace and a new temple-pyramid were left uncompleted with apparent, but perhaps deceptive, suddenness when the city was abandoned between AD 830 and 850 (Hammond et al. 1998).

The evidence for climatic as well as demographic and political causes for the end of Classic Maya civilization gives a new dynamic to our research

and leads us to seek similar environmental factors behind earlier cultural perturbations but, although these broad temporal correlations suggest that climate, climate change, and possible severe climate stress played a role in Maya cultural evolution, the chronological uncertainties imposed by relatively coarse archaeological chronologies based on ceramic sequences and calibrated radiocarbon dates preclude detailed correlation of climate changes with major archaeologically documented cultural transformations (Hodell et al. 2007; Yaeger and Hodell 2008). Despite more and better empirical data from a widening range of sources and scientific techniques, the Classic Maya Collapse remains, and will continue to remain, fertile ground for research and continuing controversy.

References

Adams, R. E. W. 1973. The collapse of Maya civilization: a review of previous theories. In *The Classic Maya Collapse*, ed. T. P. Culbert, 21–34. Albuquerque: University of New Mexico Press.

Aimers, J. J. 2007. What Maya Collapse? Terminal Classic variation in the Maya lowlands. *Journal of Archaeological Research* 15: 329–77.

Cook, O. F. 1921. Milpa agriculture: A primitive tropical system. In *Annual Report of the Smithsonian Institution* 11: 307–26. Washington, D.C.: Smithsonian Institution.

Cooke, C. W. 1931. Why the Mayan cities of the Peten District, Guatemala, were abandoned. *Journal of the Washington Academy of Sciences* 13: 283–87.

Culbert, T. P., ed. 1973. *The Classic Maya Collapse*. Albuquerque: University of New Mexico Press.

Culbert, T. P., and D. S. Rice, eds. 1990. *Precolumbian population history in the Maya lowlands*. Albuquerque: University of New Mexico Press.

Demarest, A. A. 1997. The Vanderbilt Petexbatun Regional Archaeological Project 1889–1914: overview, history, and major results of a multidisciplinary study of the Classic Maya collapse. *Ancient Mesoamerica* 8: 209–27.

Demarest, A. A., P. M. Rice, and D. S. Rice, eds. 2004. *The Terminal Classic in the Maya lowlands: collapse, transition and transformation*. Boulder: University Press of Colorado.

Diamond, J. 2005. *Collapse: how societies choose to fail or succeed*. New York: Viking.

Gill, R. B. 2000. *The great Maya droughts*. Austin: University of Texas Press.

Gill, R. B., P. A. Mayewski, J. Nyberg, G. C. Haug, and L. C. Peterson. 2007. Drought and the Maya Collapse. *Ancient Mesoamerica* 18: 283–302.

Hamblin, R. L., and B. L. Pitcher. 1980. The Classic Maya Collapse: testing class conflict hypotheses. *American Antiquity* 45: 246–67.

Hammond, N. 1982. *Ancient Maya civilization*. Cambridge: Cambridge University Press, and New Brunswick: Rutgers University Press.

Hammond, N., G. Tourtellot, S. Donaghey, and A. Clarke. 1998. No slow dusk: Maya urban development and decline at La Milpa, Belize. *Antiquity* 72: 831–37.

Harrison, P. D. 1977. The rise of the *bajos* and the fall of the Maya. In *Social process in Maya prehistory: studies in honour of Sir Eric Thompson*, ed. N. Hammond, 469–508. London: Academic Press.

Haug, G. H., D. Günther, L. C. Peterson, D. M. Sigman, K. A. Hughen, and B. Aeschlimann. 2003. Climate and the collapse of Maya civilization. *Science* 299: 1731–35.

Hodell, D. A., J. H. Curtis, and M. Brenner. 1995. Possible role of climate in the collapse of Classic Maya civilization. *Nature* 375: 31–34.

Hodell, D. A., M. Brenner, J. H. Curtis, R. Medina-González, E. Idelfonso-Chan Can, A. Albornaz-Paz, and T. P. Guilderson. 2005. Climate change on the Yucatan Peninsula during the Little Ice Age. *Quaternary Research* 63: 109–21.

Hodell, D. A., M. Brenner, J. H. Curtis, R. Medina-González, E. Idelfonso-Chan Can, A. Albornaz-Paz, and T. P. Guilderson. 2007. Climate and cultural history of the Northeastern Yucatan Peninsula, Quintana Roo, Mexico. *Climate Change* 83: 215–40.

Martin, S., and N. Grube. 2000. *Chronicle of the Maya kings and queens.* London: Thames & Hudson.

Peterson, L. C., and G. H. Haug. 2005. Climate and the collapse of Maya civilization. *American Scientist* 93: 322–29.

Sabloff, J. A. 1973. Major themes in the past hypotheses of the Maya Collapse. In *The Classic Maya Collapse*, ed. T. P. Culbert, 35–40. Albuquerque: University of New Mexico Press.

———. 1990. *The new archaeology and the ancient Maya.* New York: Scientific American Library.

Sabloff, J. A., and G. R. Willey. 1967. The collapse of Maya civilization in the southern lowlands: a consideration of history and process. *Southwestern Journal of Anthropology* 23: 311–36.

Sharer, R. J. 2006. *The Ancient Maya* (6th ed., with L. P. Traxler). Stanford: Stanford University Press.

Stephens, J. L. 1841. *Incidents of travel in Central America, Chiapas, and Yucatan,* 2 vols. New York: Harper and Sons.

Webster, D. L. 2002. *The fall of the Ancient Maya.* London: Thames & Hudson.

Wilk, R. R. 1985. The Ancient Maya and the political present. *Journal of Anthropological Research* 41: 307–26.

Willey, G. R., and D. B. Shimkin. 1973. The Classic Maya Collapse: a summary view. In *The Classic Maya Collapse*, ed. T. P. Culbert, 457–502. Albuquerque: University of New Mexico Press.

Yaeger, J., and D. A. Hodell. 2008. The collapse of Maya civilization: assessing the interaction of culture, climate, and environment. In *El Niño, catastrophism, and culture change in ancient America*, ed. D. H. Sandweiss and J. Quilter, 187–242. Washington, D.C.: Dumbarton Oaks Research Library and Collections.

11

Climate Crises and the Population History of Southeast Asia

MARU MORMINA AND CHARLES HIGHAM

INTRODUCTION

Mainland and Island Southeast Asia (MSEA and ISEA) occupy a pivotal position in any consideration of the expansion of anatomically modern humans (AMH) from Africa, and the subsequent history of those populations. Although different models are still being discussed, archaeological and genetic data are converging towards a consensus view on the early patterns of settlement by AMHs following their expansion out of Africa. According to this view, SEA is one of the final outposts of a single coastal dispersal through the rim of the Indian Ocean into Australia (Stringer 2000; Walter, Buffler et al. 2000; Field and Mirazon Lahr 2005; Macaulay, Hill et al. 2005; Sun, Kong et al. 2005; Thangaraj, Chaubey et al. 2005).

However, despite its key geographic location, little is known about the palaeodemography of SEA. Although several methodological issues still await clarification, the earliest archaeological record suggests that the area was occupied by ~45,000 years before present (YBP) (Barker, Reynolds et al. 2005), and genetic evidence suggests an even earlier date around 60,000 (Macaulay, Hill et al. 2005; Thangaraj, Sridhar et al. 2005). The fate of these early populations and their relative contributions to the present ethnic composition of SEA is a matter of debate. While some argue for biological continuity of modern SEAs with their Pleistocene ancestors (Meacham 1984–85; Bulbeck and Lauer 2006; Solheim 2006; Turner 2006), others propose a much more shallow ancestry in the Neolithic populations of East China who resettled the area in the mid- to late Holocene (Higham 1996; Higham 2002; Higham 2002; Matsumura and Hudson 2005; Lertrit, Poolsuwan et al. 2008; Matsumura, Oxenham et al. 2008).

This polarization somehow fails to recognize the paramount importance of environmental and climatic changes in shaping the demographic tapestry of SEA. This is well documented in the biogeography of the area, where geographic isolation and endemism resulting from the process of island formation shaped the distribution of flora and fauna. The extreme case of *Homo*

floresiensis suggests that these unique palaeoecological processes have also shaped the evolutionary history of *Homo sapiens*.

CLIMATE CHANGE AND HUMAN ADAPTATION IN SOUTHEAST ASIA: THE ARCHAEOLOGICAL EVIDENCE

The past 20,000 years have been characterized by a series of environmental and climatic changes that profoundly influenced this region. While the climate in SEA itself remained relatively warm and inviting to human settlement, the melting of the late Pleistocene ice in the more extreme latitudes led to a sharp rise in sea level. Nowhere was so much land gained or lost as a consequence of sea level changes than in SEA (Fig. 1). The seas that now enclose ISEA are uniformly shallow. A fall in sea level of 120 m, as occurred at the height of the Late Glacial at 21,000 years ago, would have created a massive low-lying continent larger than India or Europe (Fig. 2). Then, great rivers flowed through Sundaland, and an open savannah corridor would have eased communication and movement (Fig. 3) (Bird, Taylor and Hunt 2005). The coast of China was 600 km offshore and linked the mainland with Taiwan. Huge inland seas lay between Borneo and the Philippines, and between Sulawesi and Borneo.

With the end of the last Ice Age, global warming induced major deglatiation events leading to rapid episodes of sea-level rise, particularly during the meltwater pulse (MWP) 1A around 14 YBP (Hanebuth, Stattegger et al. 2000). These episodes resulted in the drowning of much of Sundaland; the formation of the islands of Borneo, Sumatra, Java, and many others; and a doubling of the length of the coastline. This rise did not cease at the present shore but, from 6000 to 4500 BP, the sea transgressed many tens of km inland, forming a series of raised beaches punctuated by river estuaries that sustained a thick belt of mangroves.

Who was living in SEA during this period of climatic crisis? We have known for nearly a century that inland rock shelters were occupied by small groups of mobile hunter-gatherers, whose occupation remains are very similar to those left by the Sakai, a surviving group living in Peninsular Thailand. At Tham Lod and Ban Rai, Shoocondej (2004, 2006) has uncovered not only the living floors and middens of these so-called Hoabinhian people, but also rare human burials, the dead being interred in flexed positions with minimal grave offerings. In Peninsular Thailand, at Moh Khiew, Pookajorn (1994) has recovered evidence not only for typical Hoabinhian stone tools but also four inhumation burials, one of which may have been interred in a seated, crouched position. A radiocarbon date suggests that one of the burials is about 25,000 years old. Lang Rongrien is a large cavern located on a limestone tower that lies between two streams. Its airy shelter has attracted human settlement over a long period and, when excavated in 1983 by Douglas Anderson, the remains of their activities were found. These included charcoal from hearths dated from at least 38,000 until 27,000 years ago (Anderson 1990).

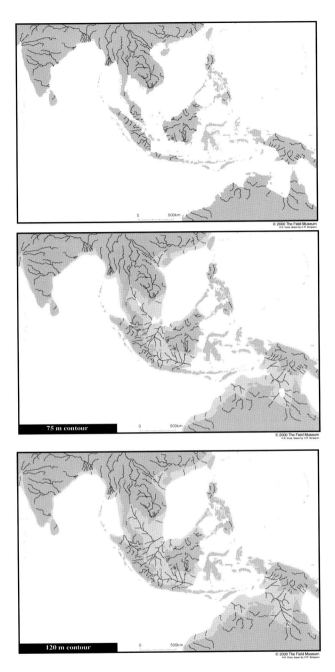

Figure 1. Changes in sea level during the last 20,000 years resulted in a massive loss of land and creation of a much longer coastline. Top, the present coastline; middle, the coastline at 75 m below modern sea level; lower, the coastline at 120 m below modern sea level (Maps courtesy of Harold Voris, Field Museum, Chicago).

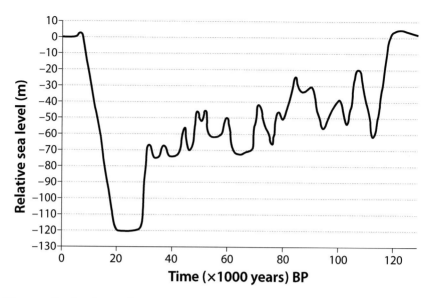

Figure 2. Sea level over the past 120,000 years.

These rock shelters by definition lie in the karstic uplands. But SEA also incorporates extensive lowland plains that have been silent until recently, in terms of early hunter-gatherer occupancy, although this terrain would have teemed with fish, shellfish, and a wide range of herd mammals. This absence of evidence is surely the result of dramatic physical changes involving deforestation and the accumulation of redeposited material over early and presumably ephemeral settlements. The presence of early hunter-gatherers in the lowland plains has now been documented at the site of Ban Non Wat in Northeast Thailand, where a group of flexed skeletons have been found dating to the second millennium BC, while a shell midden at the lowest layer in the site has yielded a radiocarbon determination of about 15,000 cal. BC (Higham and Higham 2009).

When we turn to the coast of SEA, we find that a mangrove estuary is one of the three richest habitats in terms of bioproductivity. This is partly due to the fact that mangroves lose their leaves throughout the year, thus providing the source of energy for a food chain that begins with micro-organisms and ascends through crabs, shellfish, and fish to a range of mammals, including *Homo sapiens*. For the archaeologist, the settlement history of those who lived in ancient Sundaland through the long environmental crisis that resulted from an extended period of global warming has been submerged. This loss of land through sea-level rise creates in effect a void in our knowledge that covers at least 15,000 years. In this context, it is salutary to compare this lacuna with the record we have of hunter-gatherers in Japan between 12,000 and 4,000 years ago, where a steep coastal gradient lessened the impact of the rising sea level. There, we find a vibrant and innovative sequence

incorporating fine ceramics, an impressive ritual life, and elaborate mortuary traditions.

In SEA by contrast, the only window of opportunity available to assess the adaptation to the hot, rich coastal environment lies in the raised beaches that developed when the sea transgressed beyond its present position. These sites were left stranded when, from 4500 BP, sea level fell to reach its present configuration. One such site is Nong Nor, in East Central Thailand, formerly located on a promontory that offered the sheltered conditions of a large marine embayment (Fig. 3). Similar sites are found at intervals along this shore. Excavations at Nong Nor have revealed that the occupants, about 4300 BP, lived there for at least one season of the year—the occupation remains are dominated by millions of shellfish banked up into a thick midden. Detailed examination of the midden contents reveals that these coastal hunter-gatherers ventured out to sea and brought back dolphins, eagle rays, bull and tiger sharks, and many smaller fish. They made ceramic vessels and polished stone tools. At least one person died there and was interred in a seated, crouched position, under a set of pots.

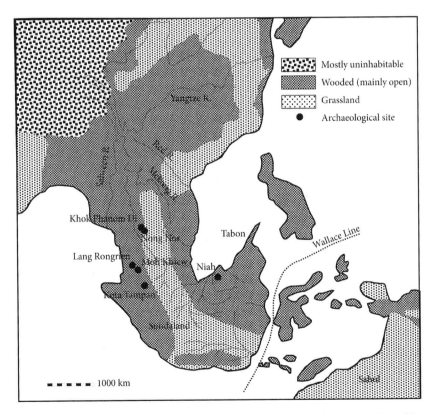

Figure 3. The coast of Southeast Asia 20,000 years ago, and the location of key archaeological sites (Shoocondej 2004).

A much larger site lies 12 km to the north. Khok Phanom Di is a site with a 500-year occupation span, and one that allows insight into the adaptation of its inhabitants to the sea-level fluctuations that so changed their environment (Fig. 4). Khok Phanom Di was occupied 300 years or so later than Nong Nor, after sea level had begun its retreat. That settlement dominated the estuary of the Bang Pakong River. The first settlers there had a material culture similar to that of Nong Nor. During the first three mortuary phases (see Fig. 5), their burials tell us that:

1. They suffered from a condition, probably thalassaemia B, which conferred resistance to malaria at the cost of suffering anaemia. Up to 40% of burials are newly born infants (Tayles 1999).
2. They made superb pottery vessels and polished stone adzes.
3. The men had powerful upper bodies, with muscular development consistent with paddling and sailing.

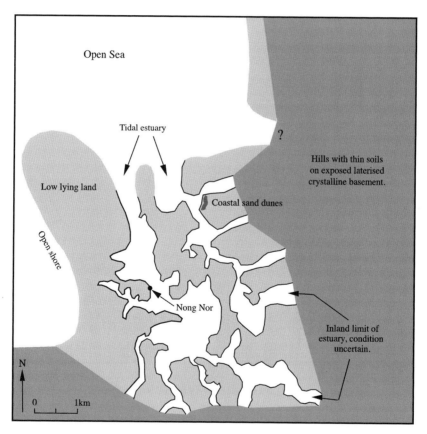

Figure 4. The reconstructed coastline in the vicinity of Nong Nor 4500 years ago. The site was located on the edge of a marine inlet (Shoocondej 2006).

4. The men had different tooth wear to women, hinting at long periods of time spent away from the site involving a different diet.

5. They were active traders.

We suggested that these findings, along with much evidence for coastal settlement in Vietnam and southern China at the same time, reflect a long and successful adaptation to the hot maritime lowlands of Sundaland. Selection for resistance to malaria, for example, would have been essential for survival, even while leading to anaemia and high infant mortality. Our inability to illuminate their ancestry archaeologically is matched by our wish to know more of the impact, on what might have been a relatively dense human population, of the climatic changes that brought on the LGM (Last Glacial Maximum), and then, through global warming, brought about a rapid series of sea-level surges. We suggest that one of the most promising lines of enquiry is through the study of modern genetic variation across this and neighboring regions. There are at least three relevant issues:

1. The first is the degree to which the descendants of the initial AMHs continued to dominate the human population throughout the period in question. Thus, the Semang of the Thai-Malay peninsula, and the Andamanese, reveal through their DNA close affiliation with the initial AMH expansion.

2. The LGM not only led to the exposure of Sundaland but also had a major impact in Central and Northern China that may well have stimulated a series of movements into SEA. These would have led to contact between the incoming groups and the local hunter-gatherer groups. Is this somehow reflected in the genetic structure of extant populations?

3. The subsequent long period of warming led to the rise in sea level and transformation of the geography of Sundaland. How did this affect the population history of the area?

HUMAN DEMOGRAPHY: A GENETIC PERSPECTIVE ON THE HUMAN RESPONSE TO CLIMATE CHANGE

At a global level, human genetic variation has been shown to be geographically structured, and several studies have suggested that geography may indeed explain better the distribution of genetic diversity than ethnicity or language (Rosenberg, Pritchard et al. 2002; Tishkoff and Kidd 2004; Prugnolle, Manica et al. 2005; Lawson Handley, Manica et al. 2007; Chiaroni, King et al. 2008). This geographic patterning can be ascribed to the effects of natural selection affecting specific adaptive loci, and genetic drift acting at a population level through migrations, founder effects, neutral variation, non-random mating, and population size fluctuations (e.g., bottlenecks) (Cavalli-Sforza, Menozzi et al. 1994). The distribution of uni-parental loci (the non-recombining part of the Y chromosome, or NRY, and the mitochondrial DNA [mtDNA]) most likely reflects the effects of genetic drift due to one or more of these factors;

hence their suitability for tracing demographic changes through time (Underhill and Kivisild 2007).

Population structure is the outcome of complex historical, social, and economic, as well as environmental circumstances. For example, major historical events, such as the conquest of the Americas or the African slave trade (Salas, Richards et al. 2004), have left a profound imprint on the genetic structure of certain groups. Socio-cultural factors such as language or ethnicity can also influence the levels of endogamy and the formation of genetic enclaves. The demography of prehistoric societies has been shown to be highly dependent on the interaction of humans with their environment. For example, palaeoenvironmental simulations suggest that, during OIS4, local environments would have played a key role in determining patterns of subsistence and mobility of AMH in their migration out of Africa through the coastal southern route into Australasia (Field and Mirazon Lahr 2005), and an open savannah corridor running through Sundaland would have facilitated human dispersals into Sahul (Bird, Taylor et al. 2005). The role of the Ice Ages has been repeatedly suggested to have impacted profoundly on the genetic structure of Western Europe, where glaciation pushed existing hunter-gatherer populations into southern refugia, and re-colonization of northern areas took place when climatic conditions improved at the beginning of the Holocene (Torroni, Bandelt et al. 2001; Achilli, Rengo et al. 2004; Forster 2004; Pereira, Richards et al. 2005). Recently, correlations between mtDNA frequencies and annual rainfall suggested that dispersals of early agricultural societies in the Fertile Crescent may have been determined by climatic fluctuations (Chiaroni, King et al. 2008).

Despite its important ecological effects, the role of climate change in the distribution of human diversity in SEA has been more or less neglected. An analysis of the genetic basis of biological adaptation to changing environments is beyond the scope of this chapter, which looks at a relatively small time window of human history, although the effects of selective pressure on human adaptive evolution in the last 10,000 years are being increasingly acknowledged (Hawks, Wang et al. 2007). Rather, we consider the demographic consequences of the human response to environmental crises, particularly sea-level rise, and its putative effects on populations adapted to coastal environments. Not only did coastlines provide ancient migration routes (Field and Mirazon Lahr 2005; Macaulay, Hill et al. 2005), they were also the main areas of settlement (Amos and Manica 2006). Water, rather than being a barrier, provided a window of opportunity for trading and colonization since ancient times. For example, maritime trade networks played an important role in the spread of Mediterranean and Indian cultures, and the colonization of the remote Pacific has long been regarded as the accomplishment of impressive marine-oriented cultures.

At the end of the LGM, coastal habitats in SEA were probably densely occupied areas, although archaeological evidence of such occupation now lies under water. Climate warming in the early Holocene, therefore, would have had a profound impact particularly on coastal dwellers, not only because of sea-level rise and the reshaping of the coastlines but also because of the ef-

fects on the distribution of flora, fauna, and overall availability of resources. Sites such as Khok Phanom Di in Thailand or Damaoshan and Tanshishan in South China indicate that coastal populations relied heavily on marine resources, even if agriculture was being practised (Jiao 2006). The archaeological evidence also suggests that during this period of sea-level rise, seafaring was developed (Jiao 2006) and trading networks around the South China Sea were probably in place (Solheim 2000, 2006).

Thus far, the only way to learn about the lives of these peoples is through what has survived in the archaeological record. Population genetic studies can provide an independent, albeit indirect, line of enquiry. In the last twenty-five years or so, a number of such studies have appeared in the literature, although SEA has remained largely unexplored. Nevertheless, a picture seems to be emerging and pointing at the impact of pre-Holocene environment on human demography.

The opposite conclusions of two recent studies on the origins of Pacific Islanders based on autosomal variation (Friedlaender, Friedlaender et al. 2008; Kayser, Lao et al. 2008) highlight the difficulties involved in using recombining loci to trace human population history, despite their potentially higher resolution. On the other hand, Y-chromosome and mtDNA loci, even though they reflect the evolutionary history of single loci, provide much clearer demographic signatures. Y-chromosome studies have suggested that the paternal heritage of most SEAs can be traced back to the Pleistocene (Capelli, Wilson et al. 2001) and seems to be indigenous. Moreover, for some of the most frequent East Asian haplogroups (C and F), genetic diversity in the mainland appears to be a subset of that in ISEA, consistent with a southern origin for most East Asian lineages (Capelli, Wilson et al. 2001). This evidence would be in agreement with a scenario of Pleistocene dispersals from ISEA into the Asian mainland that needs to be further explored.

Similarly, mtDNA studies have suggested a significant Pleistocene component in the maternal gene pool of ISEA (Richards, Oppenheimer et al. 1998; Hill, Soares et al. 2007), implying that most of the genetic diversity in the area is indigenous with only a relatively small contribution due to more recent Neolithic migrations. Genetic dating using a molecular clock suggests that about 15% of all mtDNA lineages have a deep ancestry, between 40,000 and 70,000 years (Hill, Soares et al. 2007). The geographic distribution of these lineages appears to be focused on the boundaries of ancient Sundaland, although some may spread farther north into South China. At most, about 20% of the ISEA mtDNA diversity can be traced back to Neolithic Taiwan, and many lineages seem to have spread in the opposite direction, from ISEA into Taiwan (Hill, Soares et al. 2007). These different signatures do not sit easily with a scenario of migrations from continental Asia associated with the spread of agriculture and Austronesian languages in Neolithic time. Rather, they seem to point to a demographic history of much more complex interactions that are reflected in patterns of gene flow from and into SEA. A plausible context for such complex interactions can be seen in the maritime lifestyle of prehistoric SEA populations, which still persists in some communities of maritime hunter-gatherers or sea nomads.

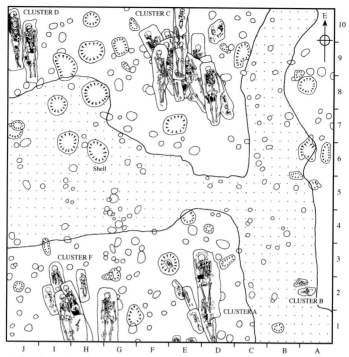

Figure 5. The burials of the third mortuary phase at Khok Phanom Di. At this stage, there appears to have been a number of immigrants into a coastal hunter-gatherer community.

Besides evidence of gene flow in the mid- to late Holocene, the maternal pool of SEA also shows an important signature of post-glacial dispersals. A number of mtDNA haplogroups seem to have spread throughout SEA and Taiwan around the time of the end of the LGM (Hill, Soares et al. 2007; Soares, Trejaut et al. 2008). The ultimate origin of these lineages is in the Asian mainland, thus suggesting that they were part of a southward migration, possibly at the height of the LGM, when populations worldwide were migrating from northern latitudes into southern tropical refugia (Forster 2004). One such lineage is haplogroup M9. This clade diversifies into two main sub-clades, M9a and E, the former exclusively restricted to the mainland and the latter to ISEA (Fig. 6). Although the origins of M9 as a whole remain somewhat elusive, this clade evolved ~50,000 years ago, probably in MSEA, where the deepest branches are found, albeit at very low frequencies. Sub-haplogroup M9a spread north into East China and Japan, probably after the end of the Ice Age, some 15,000 years ago, whereas its sister clade E most likely evolved in ISEA and was caught up in a series of range expansions from 12,000 years ago, as suggested by the strong geographic structure of its branches (Soares, Trejaut et al. 2008).

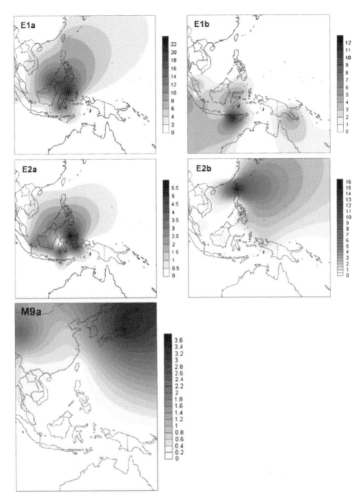

Figure 6. Spatial distribution of haplogroup M9 and its main branches M9a and E (reproduced from Soares et al. 2008).

Given the time frame of these expansions, the most likely explanation is the impact of sea-level rise and the fragmentation of Sundaland on coastal populations with marine subsistence strategies. A sudden increase in coastline would have favored the expansion of coastal-dwelling populations, and at the same time the concomitant loss of landmass would have forced inland populations with terrestrial economies to compete for resources, possibly providing strong motivation for dispersals (Soares, Trejaut et al. 2008).

Although demographic signatures are more clearly distinguishable in islands, it is possible to speculate that climate change had a similar effect on human populations in the Asian mainland. MSEA and Southern China

possibly harbor the greatest genetic diversity within East Asia, suggesting an older population history—and perhaps an ultimate origin in this area for all East Asian lineages (Jin and Su 2000; Soares, Trejaut et al. 2008)—whereas the genetic diversity in Northern China and Siberia shows a shallower time depth and is a subset of that in the south. This suggests that humans colonized northern latitudes possibly during OIS3, when moderate temperatures would have led to conditions favourable for human occupation. Subsequent cooling temperatures during the last glacial period may have led to isolation and differentiation of northern and southern Asians, and possibly to dispersals from colder regions into southern tropical areas. Some haplogroups, such as N9a, seem to have expanded into Southern China ~20,000 years ago.

The end of the LGM and the amelioration of climatic conditions provided new opportunities in terms of natural resources, ultimately favoring population growth and expansion (Scarre 2005). A recent study of mtDNA of populations in southern China suggested a signature of post-glacial expansions somewhat correlated with linguistic and ethnic diversification (Li, Cai et al. 2007). Most populations in southern China are genetically close, consistent with a process of continuous genetic admixture, despite the fact that populations remain culturally and linguistically distinct. This suggests that cultural and genetic diversity may not be totally interdependent processes, at least in this area. Thus, more gradualist models that include local cultural innovations (perhaps triggered by environmental pressures) should be considered. The evidence discussed in this chapter would seem to support such a scenario. The present genetic landscape in SEA seems more the outcome of demographic events largely shaped by human-environment interactions than the result of gene flow caused by major cultural transitions.

CONCLUSIONS

Climatic oscillations during and after the Ice Age have had a major role in shaping human population history. Perhaps nowhere in the world have these changes been more pronounced than in SEA, where sea levels rose dramatically to cover almost an entire continent, creating the world's largest archipelago. How did humans respond to such climatic upheavals? Traditionally, island fragmentation was thought to result in species endemism and isolation. Genetic and archaeological evidence in SEA suggests that seemingly destabilizing conditions, rather than posing an obstacle, provided a window of opportunity for human expansion and cultural adaptation.

REFERENCES

Achilli, A., C. Rengo, C. Magri, V. Battaglia, A. Olivieri, R. Scozzari, F. Cruciani, M. Zeviani, E. Briem, V. Carelli, P. Moral, J-M. Dugoujon, U. Roostalu, E.-L. Loogväli, T. Kivisild, H.-J. Bandelt, M. Richards, R. Villems, S. Santachiara-Benerecetti, O. Semino, and A. Torroni. 2004. The molecular

dissection of mtDNA Haplogroup H confirms that the Franco-Cantabrian Glacial Refuge was a major source for the European gene pool. *American Journal of Human Genetics* 75: 910–18.

Amos, W., and A. Manica. 2006. Global genetic positioning: evidence for early human population centers in coastal habitats. *Proceedings of the National Academy of Science* 103 (3): 820–24.

Anderson, D. D. 1990. *Lang Rongrien Rockshelter: a Pleistocene-Early Holocene archaeological site from Krabi, Southwestern Thailand.* Monograph No. 71. Philadelphia University Museum.

Barker, G., T. Reynolds, and D. Gilbertson. 2005. The human use of caves in peninsular and island southeast Asia: research themes. *Asian Perspectives* 44 (1): 1–15.

Bird, M. I., D. Taylor, and C. Hunt. 2005. Palaeoenvironments of insular southeast Asia during the Last Glacial Period: a savanna corridor in Sundaland? *Quaternary Science Reviews* 24: 2228–42.

Bulbeck, D., and A. Lauer. 2006. *Human variation and evolution in Holocene Peninsular Malaysia: bioarchaeology of Southeast Asia,* ed. M. Oxenham and N. Tayles, 133–71. Cambridge: Cambridge University Press.

Capelli, C., J. F. Wilson, M. Richards, M. P. H. Stumpf, F. Gratrix, S. Oppenheimer, P. Underhill, V. L. Pascali, T.-M. Ko, and D. B. Goldstein. 2001. A predominantly indigenous paternal heritage for the Austronesian-speaking peoples of insular southeast Asia and Oceania. *American Journal of Human Genetics* 68: 432–43.

Cavalli-Sforza, L. L., P. Menozzi, and A. Piazza. 1994. *The history and geography of human genes.* Princeton: Princeton University Press.

Chiaroni, J., R. J. King, and P. Underhill. 2008. Correlation of annual precipitation with human Y-chromosome diversity and the emergence of Neolithic agricultural and pastoral economies in the Fertile Crescent. *Antiquity* 82: 281–89.

Field, J., and M. Mirazon Lahr. 2005. Assessment of the Southern Dispersal: GIS-based analyses of potential routes at Oxygen Isotopic Stage 4. *Journal of World Prehistory* 19 (1): 1–45.

Forster, P. 2004. Ice Ages and the mitochondrial DNA chronology of human dispersals: a review. *Philosophical Transactions of the Royal Society of London B* 359: 255–64.

Friedlaender, J. S., F. R. Friedlaender, F. Reed, K. Kidd, J. Kidd, G. Chambers, R. Lea, J-H. Loo, G. Koki, J. A. Hodgson, D. A. Merriwether, and J. L. Weber. 2008. The genetic structure of Pacific Islanders. *Public Library of Science Genetics* 4 (1): 0173–90.

Hanebuth, T., K. Stattegger, and P. M. Grootes. 2000. Rapid flooding of the Sunda Shelf: a Late-Glacial sea-level record. *Science* 288: 1033–35.

Hawks, J., E. T. Wang, G. M. Cochran, H. C. Harpending, and R. K. Moyzis. 2007. Recent acceleration of human adaptive evolution. *Proceedings of the National Academy of Science* 104 (52): 20753–58.

Higham, C. 1996. A review of archaeology in mainland Southeast Asia. *Journal of Archaeological Research* 4 (1): 3–49.

Higham, C. 2002a. *Early cultures of mainland southeast Asia.* Bangkok: River Books Ltd.

Higham, C. 2002b. Languages and farming dispersals: Austroasiatic languages and rice cultivation. *Examining the language-farming dispersal hypothesis*, ed. P. Bellwood and C. Renfrew. Cambridge: Cambridge University Press.

Higham, C. F. W., and T. F. G. Higham. 2009. A new chronological framework for prehistoric Southeast Asia, based on a Bayesian model from Ban Non Wat. *Antiquity* 83: 125–44.

Hill, C., P. Soares, M. Mormina, V. Macaulay, D. Clarke, P. B. Blumbach, M. Vizuete-Forster, P. Forster, D. Bulbeck, S. Oppenheimer, and M. Richards. 2007. A mitochondrial stratigraphy for Island Southeast Asia. *American Journal of Human Genetics* 80 (1): 29–43.

Jiao, T. 2006. Environment and culture change in Neolithic Southeast China. *Antiquity* 80: 615–21.

Jin, L., and B. Su. 2000. Natives or immigrants: modern human origin in East Asia. *Nature Reviews—Genetics* 1: 126–33.

Kayser, M., O. Lao, K. Saar, S. Brauer, X. Wang, P. Nürnberg, R. J. Trent, and M. Stoneking. 2008. Genome-wide analysis indicates more Asian than Melanesian ancestry of Polynesians. *American Journal of Human Genetics* 82: 194–98.

Lawson Handley, L. J., A. Manica, J. Goudet, and F. Balloux. 2007. Going the distance: human population genetics in a clinal world. *TRENDS in Genetics* 23 (9): 432–39.

Lertrit, P., S. Poolsuwan, R. Thosarat, T. Sanpachudayan, H. Boonyarit, C. Chinpaisal, and B. Suktitipat. 2008. Genetic history of Southeast Asian populations as revealed by ancient and modern human mitochondrial DNA analysis. *American Journal of Physical Anthropology* 137: 425–40.

Li, H., X. Cai, E. R. Winograd-Cort, B. Wen, X. Cheng, Z. Qin, W. Liu, Y. Liu, S. Pan, J. Qian, C.-C. Tan, L. Jin. 2007. Mitochondrial DNA diversity and population differentiation in southern East Asia. *American Journal of Physical Anthropology* 134 (4): 481–88.

Macaulay, V., C. Hill, A. Achilli, C. Rengo, D. Clarke, W. Meehan, J. Blackburn, O. Semino, R. Scozzari, F. Cruciani, A. Taha, N. K. Shaari, J. M Raja, P. Ismail, Z. Zainuddin, W. Goodwin, D. Bulbeck, H.-J. Bandelt, S. Oppenheimer, A. Torroni, and M. Richards. 2005. Single, rapid coastal settlement of Asia revealed by analysis of complete mitochondrial genomes. *Science* 308 (5724): 1034–36.

Matsumura, H., and M. J. Hudson. 2005. Dental perspectives on the population history of Southeast Asia. *American Journal of Physical Anthropology* 127 (2): 182–209.

Matsumura, H., M. F. Oxenham, Y. Dodo, K. Domett, N. K. Thuy, N. L. Cuong, N. K. Dung, D. Huffer, and M. Yamagata. 2008. Morphometric affinity of the late Neolithic human remains from Man Bac, Ninh Binh Province, Vietnam: key skeletons with which to debate the "two layer" hypothesis. *Anthropological Science* 116 (2): 135–48.

Meacham, W. 1984–85. On the improbability of Austronesian origins in South China. *Asian Perspectives* 26: 89–106.

Pereira, L., M. Richards, A. Goios, A. Alonso, C. Albarran, O. Garcia, D. M. Behar, M. Golge, J. Hatina, L. Al-Gazali, D. G. Bradley, V. Macaulay, and

A. Amorim. 2005. High-resolution mtDNA evidence for the late-glacial resettlement of Europe from an Iberian refugium. *Genome Research* 15 (1): 19–24.

Pookajorn, S. 1994. Final report of excavations at Moh-Khiew Cave, Krabi province; Sakai Cave, Trang Province and ethnoarchaeological research of hunter-gatherer group, so called Mani or Sakai or Orang Asli at Trang Province (The Hoabinhian Research Project in Thailand). Bangkok, Silpakorn University.

Prugnolle, F., A. Manica, and F. Balloux. 2005. Geography predicts neutral genetic diversity of human populations. *Current Biology* 15 (5): R159–R160.

Richards, M., S. Oppenheimer, and B. Sykes. 1998. mtDNA suggests Polynesian origins in eastern Indonesia. *American Journal of Human Genetics* 63: 1234–36.

Rosenberg, N. A., J. K. Pritchard, J. L. Weber, H. M. Cann, K. K. Kidd, L. A. Zhivotovsky, and M. W. Feldman. 2002. Genetic structure of human populations. *Science* 298: 2381–85.

Salas, A., M. Richards, M.-V. Lareu, R. Scozzari, A. Coppa, A. Torroni, V. Macaulay, and Á Carracedo. 2004. The African diaspora: mitochondrial DNA and the Atlantic slave trade. *American Journal of Human Genetics* 74 (3): 454–65.

Scarre, C. 2005. *The world transformed: from foragers and farmers to states and empires. the human past.* London: Thames and Hudson.

Shoocondej, R. 2004. Archaeological heritage management at Ban Rai and Tham Lod Rockshelters Project (in Thai). Bangkok, Silpakon University.

Shoocondej, R. 2006. Late Pleistocene activities at the Tham Lod rockshelter in highland Bang Mapha, Mae Hongson Province, Northwestern Thailand. Uncovering Southeast Asia's Past. I. C. Glover, V. C. Pigott, and E. A. Bacus, eds. Singapore: NUS Press.

Soares, P., J. A. Trejaut, J.-H. Loo, C. Hill, M. Mormina, C.-L. Lee, Y.-M. Chen, G. Hudjashov, P. Forster, V. Macaulay, D. Bulbeck, S. Oppenheimer, M. Lin, and M. B. Richards. 2008. Climate change and postglacial human dispersals in Southeast Asia. *Molecular Biology and Evolution* 25 (6): 1209–18.

Solheim, W. G. 2000. Taiwan, Coastal South China, Northern Vietnam, and the Nusantao maritime trading network. *Journal of East Asian Archaeology* 2 (1): 273–84.

Solheim, W. G. 2006. *Archaeology and culture in Southeast Asia: unravelling the Nusantao.* Diliman, Quezon City: University of Philippines Press.

Stringer, C. 2000. Coasting out of Africa. *Nature* 406: 24–27.

Sun, C., Qing-Peng Kong, M. G. Palanichamy, S. Agrawal, H-J. Bandelt, Yong-Gang Yao, F. Khan, Chun-Ling Ahu, T. K. Chaudhuri, and Ya-Ping Zhang. 2005. The dazzling array of basal branches in the mtDNA macrohaplogroup M from India as inferred from complete genomes. *Mol Biol Evol* 23 (3): 683–90.

Tayles, N. G. 1999. The excavation of Khok Phanom Di. Volume V. *The People.* London, Society of Antiquaries of London.

Thangaraj, K., G. Chaubey, T. Kivisild, A. G. Reddy, V. K. Singh, A. A. Rasalkar, and L. Singh. 2005. Reconstructing the origin of Andaman Islanders. *Science* 308: 996.

Thangaraj, K., V. Sridhar, T. Kivisild, A. G. Reddy, G. Chaubey, V. K. Singh, S. Kaur, P. Agarawal, A. Rai, J. Gupta, C. B. Mallick, N. Kumar, T. P. Velavan, R. Suganthan, D. Udaykumar, R. Kumar, R. Mishra, A. Khan, C. Annapurna, and L. Singh. 2005. Different population histories of the Mundari- and Mon-Khmer-speaking Austro-Asiatic tribes inferred from the mtDNA 9-bp deletion/insertion polymorphism in Indian populations. *Human Genetics* 116: 507–17.

Tishkoff, S. A., and K. K. Kidd. 2004. Implications of biogeography of human populations for "race" and medicine. *Nature Genetics* 36: S21–S27.

Torroni, A., H.-J. Bandelt, V. Macaulay, M. Richards, F. Cruciani, C. Rengo, V. Martinez-Cabrera, R. Villems, T. Kivisild, E. Metspalu, J. Parik, H.-V. Tolk, K. Tambets, P. Forster, B. Karger, P. Francalacci, P. Rudan, B. Janicijevic, O. Rickards, M.-L. Savontaus, K. Huoponen, V. Laitinen, S. Koivumäki, B. Sykes, E. Hickey, A. Novelletto, P. Moral, D. Sellitto, A. Coppa, N. Al-Zaheri, A. S. Santachiara-Benerecetti, O. Semino, and R. Scozzari. 2001. A signal, from human mtDNA, of postglacial recolonization in Europe. *American Journal of Human Genetics* 69: 844–52.

Turner, C. G. 2006. Dental morphology and the population history of the Pacific Rim and Basin: commentary on Hirofumi Matsumura and Mark J. Hudson. *American Journal Physical Anthropology* 130: 455–61.

Underhill, P. A., and T. Kivisild. 2007. The use of Y chromosome and mitochondrial DNA population structure in tracing human migrations. *Annual Review of Genetics* 41: 539–64.

Walter, R. C., R. T. Buffler, H. Bruggemann, M. M. M. Guillaume, S. M. Berhe, B. Negassi, Y. Libsekal, H. Cheng, L. Edwards, R. von Coselk, D. NeÂraudeau, and M. Gagnon. 2000. Early human occupation of the Red Sea coast of Eritrea during the last interglacial. *Nature* 405: 65–69.

Index